Growing Your Property Partnership

Plans, Promotion and People

Kim Tasso

2009

Routledge
Taylor & Francis Group

LONDON AND NEW YORK

First published 2009 by Estates Gazette

Published 2014 by Routledge
2 Park Square, Milton Park, Abingdon, Oxon OX14 4RN
711 Third Avenue, New York, NY 10017, USA

Routledge is an imprint of Taylor & Francis Group, an informa business

Notices
Knowledge and best practice in this field are constantly changing. As new
research and experience broaden our understanding, changes in research
methods, professional practices, or medical treatment may become necessary.

To the fullest extent of the law, neither the Publisher nor the authors, contributors,
or editors, assume any liability for any injury and/or damage to persons or
property as a matter of products liability, negligence or otherwise,
or from any use or operation of any methods, products, instructions, or ideas
contained in the material herein.

ISBN   978-07282-0553-6

Typeset in Palatino 10/12 by Amy Boyle, Rochester
Cover design by Rebecca Caro

Contents

PROMOTION

PEOPLE

Dedication

Having dedicated previous books to my father and my children, on this occasion I would like to acknowledge my best friends for sharing the cheers, leers and occasional tears. Here's to you guys: Darryll Adler, Roger Bell, Mark Brandon, Lisa Coletti, Elaine Crittenden, Gerry Dawson, Julie Hobbs, Malcolm Lewis, Julie Moment, Anita Paskin, Melanie Robinson, Hazel Rosemin, George Theo and, of course, my best friend Caroline Matthew and my ex-husband William Tasso.

Acknowledgements

There are so many people to thank — where do I start?

A starting point is in 1999 when, after several years working with accountants and lawyers, I began three very happy years with a fantastic partnership called Weatherall Green & Smith. I continue to thank Terry Knight and Greg Cooke and the other partners there for their indulgence and support during that time and I loved the journey we made to create the new Atisreal. Since then I have worked with many clients in the property sector and I owe them my thanks for their trust in me. Some of these clients — along with some of the friends I have made along the way — deserve special thanks for contributing case studies to this book:

- Brown & Co
- Chase Buchanan
- Chase & Partners
- Cluttons
- Coda Studios
- Colliers CRE
- Douglas and Gordon
- Drivers Jonas
- Henry Adams
- Pellings
- RICS
- Spacelab
- Stiles Harold Williams

I must also thank my long-suffering publishers — EG Books — and the staff who put up with the inevitable delays and tantrums in producing any publication. Thank you so much Paul Sayers and Sarah Jackman for your encouragement, patience and support. Also to Adam Tinworth, Peter Bill and Alison Bird, who were my original contacts at *Estates Gazette* more years ago than I suspect any of us wish to remember.

Particular thanks are due to those who provided specialist input. The clever stuff is down to them. Any mistakes are mine. Thanks for technology input from Andrew Waller of Remit Consulting (and my apologies for not including the additional chapter he so carefully prepared for which there was sadly insufficient space in the end — but he assures me that if any readers get in contact with him he will be happy to email them a copy) and, on the financial front, it is major thanks to George Crowther from haysmacintyre accountants.

Of course, I must thank those who continue to support me day to day — my wonderful children James and Lizzie, of whom I am immensely proud, my father Ken, who, as a quantity surveyor, was the spark for my original interest in the property industry, my brother Neil and his family and my dear friends to whom I have dedicated this book.

Introduction

Why a management book for surveyors and agents?

There are two reasons for writing this book: the first is personal and the second is professional.

The personal reason first. Several years ago, I was coaching a senior surveyor who had been promoted from head of his division to become managing partner of his firm and responsible for a fast-growing £50m business. He was incredibly bright and had plenty of commercial common sense. Yet he had never had any formal management training and he reacted strongly to any suggestion of theory or jargon.

As I structured his coaching programme, I talked about the various books that he might read to help him quickly acquire the fundamental building blocks of management. But his wry response was "But Kim, I only read one book a year".

So for brilliant surveyors who have little time or appetite for reading, my aim is ambitious — it is to provide that one book that a "reluctant-to-leave-client-work-to-do-management" surveyor or agent might read that will provide the foundations of management knowledge to support him (or her) in a management role — whether it is for a small team or a medium-sized firm. I aim to avoid most of the business and management jargon and to make it as painless and as quick a read as possible. I have also tried to adopt the slightly irreverent and "let's not take ourselves too seriously" tone that is so common among the largely modest property professions.

My second reason is professional. I have been working with professional practices for more than 20 years — lawyers and accountants as well as surveyors. I have to say that the lawyers and accountants seem to have grasped the management concept and moved from the partnership ethos to corporate culture much faster than surveyors. And they have reaped the benefits — with phenomenal growth rates and eye-watering profits. I would like to help surveyors catch up a little and perhaps regain some of the lost ground — and to win back their place at the client boardroom table alongside the accountants and lawyers. Particularly after the traumatic effects of the financial and property market meltdown during 2008 — it's got to get better during 2009!

I find it odd that despite all the training that surveyors must complete to achieve their professional accreditation, to survive they must know how to manage people, generate business and keep clients happy yet these "soft" management subjects are not an integral part of their professional training. To fill this gap, I have tried to provide a simple and palatable introduction to these subjects but in a way that is directly relevant to the challenges of today's property partnerships. My aim is to be pragmatic and to offer advice and suggestions for tackling the real issues that you must face each day as you struggle to balance your role as busy fee-earner or deal maker with that of manager of a team or firm.

How to use the book

You could read this book from start to finish if you wanted to gain a general overview of the main management ideas. Apart from finance and IT (and managing the professional work and transactions that make up your daily professional life) — which are, of course, fundamental for the success of any business — most other management subjects are covered. Albeit briefly. There is a huge wealth of material on management, leadership, marketing, selling, human resources and change management — so I have had to be selective. I made some hard choices — and no doubt made some mistakes along the way with some serious omissions — but I based my decision on what I have used and found to be of most value in the property sector in the course of my consultancy work and journalism.

A word of warning. Planning is one of those topics that the property professions find amongst the hardest. This is because it involves a lot of thinking, and property people prefer to do stuff. While planning is important — and is the first chapter in the book — it is probably the toughest and hardest going read in the book. So if you read it first, do not be put off the rest of the book. Once you have read it, the rest of the ride will be fun — and easier.

If you have suddenly found yourself within a management role — such as the surveyor mentioned above — then it would be prudent for you to read the entire book. That way you will be able to gain a better picture of the task ahead and obtain a framework of setting priorities for your first year of management. You will need a friend! You might also refer to *Practice Management guidelines for surveyors* RICS 1997. It contains some more great checklists — especially if you are looking at quality accreditation for something like ISO 9000. However, this book tries to answer some of the many questions posed in there and offers some pragmatic solutions to common challenges and questions.

I suspect that the book will be used more as a reference that surveyors and agents might "dip into" when they are faced with a particular challenge or situation or when they are seeking some fresh perspective on a reoccurring problem in their practice.

Periodically throughout the chapters there are suggested checklists — and this should help direct you towards the necessary things you need to do in order to make progress on a particular issue and move your practice towards modern management. If you have not had time to prepare for a particular board or management committee meeting, these lists will also provide you with an instant agenda.

The case studies — for which I am extremely grateful to the contributors — offer an insight into how real firms have tackled some of the core management challenges. The research we conducted before writing the book indicated that while small and medium firms wanted to know what other small and medium-sized firms had done, they also wanted to know what the largest and most prestigious firms were doing. Therefore, I have included case studies from both. But hopefully I will continue to write in *Estates Gazette* about all these issues, so if you would like to contribute a case study for the future — or to give your views on the advice and ideas in this book — then please get in touch.

Planning the future shape of your firm (planning strategies)

Roadmap

> *"If you fail to plan, you plan to fail."*
> Philip Kotler

> *"Plans are only good intentions unless they immediately degenerate into hard work."*
> Peter F Drucker

When I ask property partnerships if they have a business plan they usually answer: "Of course" and give me a rather old-fashioned look of disapproval. The plan is to grow the business and become rich(er)! But when I ask them to describe to me how their firm will look in three or four years' time, or what the major changes in the market will be and how they plan to respond to them I receive a less confident response.

This is because what they *think* is a business plan is really a production plan of what is likely to happen during the next year — a budget that shows the relationship between the number of people working at the firm, the number of deals they will do (or hours they will work) and how much income this will convert into. Often the income is based on an expected increase of somewhere between 5% and 12%. Then it shows the likely expenditure and — the figure that everyone is interested in — the anticipated profit. So the business plan is, at best, a financial budget and, at worst, wishful partner thinking.

So what you have isn't really a business plan is it?

This chapter is divided into two parts. First, I need to convince you and your partners that it is a worthwhile investment of your time to tackle — head on — the core issues at the heart of your partnership. There is no point embarking on a business planning process unless you are really committed to doing so and then have the buy-in from your partners. Second, I need to provide a pragmatic and pain-free step-by-step guide on how to produce a workable (not MBA class level) business plan that suits you and your partners.

10 reasons why you need a proper business plan

Sometimes partnership or board meetings become repetitive. These are some of the most common reasons:

1. Without an agreed and clear direction, every small decision involves a long and detailed discussion that goes over the same ground — again!

2. The same issues come up time and time again and are never properly addressed — because we are trying to avoid conflict and there are some potentially big differences about what different partners want and no one wants to start a fight.

3.   No one is taking any action to address matters — partners are great at having ideas, but there is no time to make any of it actually happen.

While point 3 can be addressed by forming the appropriate management structure, all these issues can be eased by the development of a business plan that sets out the firm's broad direction once and for all. The business plan — and overall strategy — then tends to dictate the main focus areas of future partnership or board meetings and also provides answers for many issues and questions before they even turn up.

Here is an example. Let's say partners keep nominating different firms for merger talks. Through the process of developing a business plan the partners will decide on whether they want a merger or not, the likely scale and timing of such a merger, the likely location, the profitability profile of potential targets, the key values that must be held by the other firm as well as the person in the firm responsible for assessing mergers and even the priorities.

Another example? Let's say one particular partner keeps requesting additional resources to develop his (or her) team. The overall business plan will have considered all the elements of the firm and identified the levels of growth required from each and also, therefore, the corresponding resources available to each team. This ends the constant debates about further resources.

So why do you need a proper business plan? Well, there are many reasons. I won't bore you with them all, but here are some of the more important reasons:

1.   You need to have some *agreement between the owners* (partners) of the important things to achieve and the best ways to do this. Some may desire faster growth, some may wish for a larger team, many want more money, and some want fame.

2.   You need to *have a vision* of how you want the firm to develop — say, in five years' time — otherwise the business may grow in unexpected ways. If you remain preoccupied with doing the work you have today, your practice will drift. It may even drift in the wrong direction, so you find yourself with a practice focused on the wrong sort of clients and the wrong sort of (unprofitable) work.

3.   You need to take a systematic look at your business and also at the *external market* to check that your aims, assumptions and plans are

realistic. As I write, the property market is in a bad state, with residential sales almost at a standstill and droves being made redundant from the large commercial firms. Yet a cursory glance at the figures would have given ample warning about the impending slowdown and smart firms could have made suitable adjustments to their plans, staffing, services and positioning to have made hay while the rain fell on its competitors.

4. *Things change.* Markets. Clients. Technology. Laws. The aspirations of young professionals. Even quite a small change in any of these areas can provoke a huge change in the demand for your services — or your ability to meet those needs — and thus the future success of your firm. Getting into the habit of constant vigilance and creative thinking on how to grasp the opportunities and side step the threats means that you will proactively adapt and change so that you are not caught unaware and facing financial disaster. There are stories about firms that are myopic — that see only a very narrow slice of the market and fail to see the impact of broader changes hurtling towards them. For example, who anticipated the impact of the online revolution on residential property sales? Who was ready for the obsession with sustainability?

5. An essential part of the planning process is careful research and analysis so that you can *anticipate future needs* and make the necessary investment and changes to your practice so that you are well positioned to take advantage of, rather than be at the mercy of, new market and client demands. Otherwise, the danger is that you will be constantly on the back foot, trying to assess and respond to a barrage of changes that are competing for your management time and scarce investment resources.

6. With a realistic plan of where you are, where you want to get to and how you will get there you will be able to *persuade other firms and senior people* to join you. Too often merger discussions or negotiations with senior lateral hires flounder because the partners fail to impress with a clear commitment to a known future and a carefully prepared plan of what is required to achieve that future. Clients and existing staff will be reassured too.

7. As I mentioned above, without a good plan you will *waste a lot of time* in management meetings talking around the same old topics

again and again. The planning process will raise new issues and topics to discuss and help you break out of old ways of thinking. It means that you can assess all the options and alternatives together and make a considered choice. With a clear plan it is much easier to deal with each new opportunity or threat as it arises.

8. When you know where you want to get to, you know what you have to do. Sometimes the work involved in getting there *involves large, expensive, high-risk and complex projects* that someone has to manage. It will be easier to agree who is doing what and by when — and how they will be measured — when there is a clear plan. Furthermore, showing the strategic importance of these big "hot potato" projects will provide the motivation to the poor folk who take them on.

9. There is *only limited time and money* to invest in growing the firm for the future, rather than financing today's operations. A proper plan will illustrate what needs to be done and why, and the importance of each programme — whether this is in HR, office premises, marketing, technology or some other area of activity.

10. All professional practices are reliant on their staff for their short and long-term success. Sharing the owners' vision and plan with all members of the firm will build confidence, boost morale, *build team spirit and motivation* and mobilise them into helping you make your vision a reality. It is a key element in staff retention.

What is a business plan?

As I mentioned above, many firms will have a plan that considers next year's budget. This is usually developed on the basis of a "bottom up" analysis of what each surveyor/agent or team plans to do. Figure 1.1 shows this process.

While this gives an idea of the desired future income — and the required levels of staff — it is a rather internally focused exercise, which makes the big assumption that life will continue in much the same way that it has in the past. Let us remember 2008 as the year when everything changed. Most of the team plans will be based on an assumption of a reasonable amount of growth — say 5–10% — without really considering whether this is realistic. It also means that

Figure 1.1: Bottom-up planning

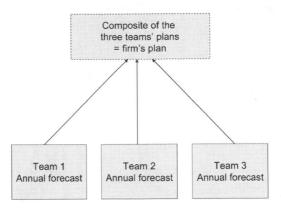

Source: Kim Tasso

if some teams anticipate greater growth than others — and therefore demand more resources — the firm may start to grow into a different shape than before. So you may have the following situation occurring:

Figure 1.2: The changing shape of your practice

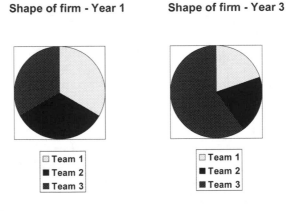

Source: Kim Tasso

Now this is all fine and dandy — and some partners may be of the view that all growth is good growth. But the team experiencing the fastest growth may be doing work that is less profitable than that done by the other teams. Or it may be serving clients in a rather different market to the other teams. Or using resources that could be used to build new teams or support more profitable growth in the other areas. Or it may skew the firm's reputation and profile towards a particular market or specialism — making it even harder for the other teams to grow down the line.

Many firms believe that it is important to allow their partners a high degree of autonomy so that they can pursue the opportunities in their markets. However, as the firm grows and it grapples with managing limited resources — whether this is qualified surveyors and agents, skilled support staff, technology resources or marketing budget — it must start to make choices about which opportunities the firm wishes to pursue with those limited resources.

Your business plan needs what we call "an umbrella plan" where the partners agree together what the overall aims of the firm might be. The individual teams build their plans using the overall aims and parameters in that central plan. So we have a top-down and a bottom up approach to the business planning process:

Figure 1.3: Umbrella plan — combining top-down and bottom-up planning

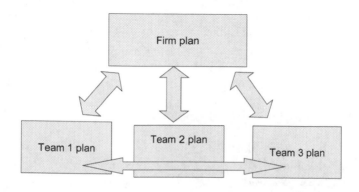

Source: Kim Tasso

In a nutshell, your business plan guides you through a simple process that answers the following questions:

- Where are we now?
- Where do we want to be?
- How will we get there?

These three harmless looking questions are a little tougher to answer than they appear at first glance. To answer properly — where are we now? — we need to gather and examine a range of information from outside and inside the firm. Then we have to get all of the owners of the business — and possibly some of the future owners of the business — to agree the overall purpose of the business and some measurable goals to aim for. There are always lots of different ways to skin a cat, so the choice of different strategies for how you reach those goals is likely to result in some healthy and sometimes heated discussions.

Seven-step guide to producing a business plan

Here is a simple seven-step guide to producing your plan. Basically, we are going to follow these stages:

1. *Set preliminary objectives*
    Sometimes firms wait until they have completed the next stage of analysis before setting goals, others will have goals from previous business plans or have an idea at the outset of what they want to achieve.

2. *Undertake external and environmental analysis*
    A systematic review of what is going on outside the firm — both close to home, such as competitor action and client market trends and longer term and more distant, such as the economy, technology trends and the political scenario.

3. *Conduct a thorough review of the internal situation and your resources*
    Take a magnifying glass around your firm and look at its finances, technology, staff, services, marketing and every other aspect of its performance.

4. *Summarise the analyses*

    From the internal analysis you will have identified some strengths and weaknesses, and the external analysis will have revealed opportunities and threats. Essentially, the business planning process is about matching what is happening within your firm to the external environment. Think of it as a reality test.

5. *Review objectives and consider the alternatives and options*

    Now you can consider what you want to achieve in the context of the internal and external reality. You will know what the options are — grow, shrink, focus — and there will be numerous ways to achieve these options. Therefore, you need to assess each option against your aims and the internal and external reality.

6. *Decision making*

    Now the partnership must concentrate on making decisions between those various options and alternatives. These decisions will be shaped by the partnership's values and culture — and by its attitude towards what they now call "corporate social responsibility" — just how much you want to take out and how much you want to put back in — to your staff, to the future, to the community, to the profession and so on.

7. *Revise objectives and establish monitoring systems*

    Now that the decisions are made you may need to adjust your initial objectives. Large investment now may mean less profit today but more for tomorrow. You also need to agree how often you will review the plan and progress and what you (and others) will do if you deviate from the plan.

Here is a template, or series of standard headings, that you should have in your business plan. The remainder of this chapter offers some help in trying to complete the plan.

The simple business plan outline
Executive summary
Vision and mission

Objectives   Fees and profits
                                       Growth, shape and size

Strategic choices   Markets and clients
                                       Products and services

Strategies — internal and infrastructure	Investment
	People
	Process and technology
	Premises and facilities
Strategies — external	Quality, client care and service
	Marketing and selling
	Relationship management

That's not so scary but the devil is in the detail and what appear to be relatively simple topics can require a lot of research, analysis and discussion in order to identify the wheat and dispel the chaff.

Checklist and actions

- Obtain a copy of the previous or present business plan. Is it a real plan or simply a statement of the things on the partners' minds or short-term actions?

- Consider the issues raised at the next partners' meeting — strategic and long term or operational and short term?

- Obtain your partners' agreement that there is a need for a proper business plan.

- Make sure your partners understand what is involved in the planning process.

- Consider whether you generally adopt a bottom-up or a top-down approach to business planning.

- What is the overall umbrella strategy of your practice?

- Think about which partners (and associates) might form a small group of motivated people to tackle the business planning task.

- Develop a timescale for the research and preparation tasks and schedule a meeting for when the first draft can be discussed.

But things are changing too fast for a plan

This is a common objection to devoting the necessary time and resources to business planning. It is a valid viewpoint in some regards. However, remember the quote at the start of the chapter — "failing to plan is planning to fail". In business planning, the journey is as important as the destination. You will learn a lot from the research and be better prepared to tackle the changes identified and those that suddenly emerge in a comfortable time frame — rather than reacting to wave after wave of market changes. How would the property industry have fared if it had been more alert to the imminent recent financial meltdown and credit crunch?

The other thing to realise is that in the real world only some of your planned strategy will materialise. All sorts of other opportunities and threats will emerge and some of your partners will take these in their stride and deal with them. So in reality, the business plan is only part of the equation.

Figure 1.4: Intended and realised strategy

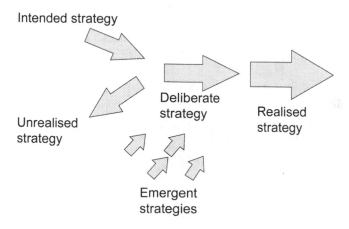

Source: Mintzberg and Waters

While emergent strategies are good, there should be some intended strategy to temper this — otherwise your practice may "drift" into unintended markets and positions.

Your mission — if you choose to accept it

Just before we get into the rigorous process of producing a business plan, I would like you and your partners to step back and enter the realms of "why are we here?".

Some firms will not be shy in telling me: "We just want to make a lot of money as fast as possible". This is a rather different mission to the one that says: "We want to have a pleasant work life balance and make a reasonable living", "We want to be the leaders in our field within five years", "We want to merge with another firm and get at least two million apiece and exit within five years" or even "We just want to do great work for great clients, make a reasonable living and pass on a healthy practice to the new generation of partners who will support our pensions".

While these are the personal missions of the owners of the business, we need to think about the collective mission of the partners and how it is expressed day to day in your business operations.

A useful definition of the mission statement was provided by Abell:

1. The client groups that will be served — what markets and groups or segments are we concentrating on?
2. The client needs that will be met — what exactly are the needs that we will be meeting? What value do we bring to our clients?
3. The "technology"/ways that we will satisfy those needs — is it through a great knowledge base, highly experienced people, clever systems, deep relationships with our clients? What is it that we do so well that people buy our services?

Factors to be taken into account in a mission statement:

1. The firm's history and its performance.
2. The preferences, values and expectations of the partners and others who have power.
3. Environmental factors — what's going on in the real world that may affect our market.
4. The resources available — the skills, people, technology, investment.
5. Distinctive competencies — what is our firm particularly good at?

It must also refer to the competition:

1. Industry scope — are we generalists or specialists? Are we in the

commercial or residential market? Are we supporting all types of buildings or one particular type?

2. Geographical scope — are we aiming at global, national, regional or local markets?
3. Market segment scope — do our clients come from all sectors or are there some particular markets of interest? Where are we concentrating our efforts?
4. Vertical scope — what will we do ourselves and what will we leave to others? What specialist services will be buy in? Where will we collaborate with others?

Jim Collins, in his book *From good to great*, used a visual of a hedgehog to suggest that great firms had a BHAG (Big Hairy Audacious Goal) and that you developed it by considering:

1. What are you deeply passionate about?
2. What drives your economic engine?
3. What can you be the best in the world at?

I worked with a 200-partner property practice a while ago that had worked hard at developing their BHAG and it drove just about everything they did. Every member of staff knew and was committed to the BHAG and they achieved incredible (client, income and profit) growth as a result.

Sometimes you need to go through the detailed audit process (see below) before you have enough information to formulate a draft mission statement that can then be discussed. Here are some examples of property firms' mission and value statements that I found on their websites:

Drivers Jonas, what we're about:
1. Putting our clients first.
2. The quality of our people.
3. Innovation and creativity in everything we do.
4. A culture of openness and sharing.
5. Responsibility to the environment.
6. Engagement with communities.

Workspace Group plc is a specialised property-based business devoted to providing office, studio and light industrial workspace for small businesses. The parent company and its subsidiaries assist new and existing small businesses by creating affordable accommodation for rent on flexible and user-friendly terms. The Group aims to achieve an acceptable financial return for its shareholders while at the same time providing a service to small business communities and encouraging urban regeneration and co-operation between the private and public sectors. Workspace Group plc understands that the best interests of its shareholders are served by actions that recognise its shared interests with customers, employees and the wider community.

Pellings: We will build on and improve our reputation for providing multi-disciplinary property and construction consultancy services in a professional, open and unique way, giving added value to our clients whilst offering a positive and supportive environment for all staff.

Davis Langdon: Our mission is to provide managed solutions, which reduce risk and maximise value, for clients investing in infrastructure, construction and property.

Fitzpatrick Property Consultants: Passionate about property and recognise that all buildings are unique and contain their own characteristics, challenges and solutions. Integrity and honesty are the foundation of our operations combined with a meticulous and conscientious approach supported by a firm but fair method of working. Our culture embodies the highest levels of professionalism, accountability and excellence. Fitzpatrick Property Consultants listen to our client's specific requirements, understand their objectives and aspirations and deliver to the highest possible standards of best practice. Our mission statement is "To leverage greater value from our client's property assets and consistently exceed our client's expectations".

Rightmove: The company's objective is to grow its customer base and revenues by connecting people with property. The company is pursuing the following strategy to achieve this objective: Defending the clear market leadership of the Rightmove website, new service offerings and selective acquisitions in support of the strategy.

The British Council for Offices' mission is to research, develop and communicate best practice in all aspects of the office sector. It delivers this by providing a forum for the discussion and debate of relevant issues.

Rok aims to become the Nation's Local Builder™ by operating from a network of offices in major towns across the UK, using dedicated teams who live and work in their communities.

Some suggest that the mission statement flows directly from the vision statement. "What's that?" I hear you ask. The vision statement is along the lines of a short "where we want to be" statement so the mission statement says how you will implement and realise the vision — a "how will we get there" response.

A vision is an idealised scenario of what is to come — a goal that entices people to work towards an exciting future. Just jot down in a few words what you want to achieve and how you are going to do it. Blend the reality of today with the aspiration of tomorrow. Hopefully, you will end up with a simple statement that generates an emotional and motivational response, is easily understood, influences action and is realistic and believable.

Developing a business plan

Let's look forward to getting our brains working in a way that maybe they have not done before and to think about what we do and how we do it and how we might need to change. If you need some more motivation, think about how much you will save in consultant fees from people like me if you spend a bit of time doing the business plan yourself rather than getting someone else to do it for you!

Think of the business plan as a journey of discovery. The real value in the business planning process is that it helps everyone gain a better understanding of the overall shape, strengths and weaknesses of the business and its market(s), the challenges ahead, the strategic alternatives and the work that must be done to achieve the agreed goals.

Don't let the committee design the camel

So what do you need to do to develop a business plan and is it going to take all of your time? Yes, it can take some time, but it doesn't have to and you should spread the load a bit by getting others involved to help you.

The more people you get involved in the business plan development process, the more people you will have with a greater understanding of the challenges ahead and prepared to commit to helping to achieve the goals. Ideally, you should have a small team of senior people to drive the business planning process — drafting in others to assist with specific analysis or thinking tasks as you go along.

While you need all owners (partners) to be involved in the process of developing the plan — and certainly in terms of setting the broad aims for the business over the next three to five years — you cannot design the entire business plan by committee.

Either assign the task to up to three individuals or divide various parts of the plan and allocate them to different partners to compile a first draft. Then get two or three people (or just the managing and senior partner) to do a final edit of the document before it is circulated to all the partners for discussion and approval.

The basic stages are to undertake some detailed analysis both of what is going on inside your firm (resource analysis) and outside your firm (environmental analysis). Then you need to set some objectives (although you may take a little detour to decide on your vision and mission first — then fill in the detailed objectives afterwards). It is then time to look at the various alternative courses of action available to the firm and to make some decisions about what to focus on in the short to medium term (as well as in the longer term). At some point, you need also to consider the partnership's attitude to risk — some firms are very conservative while others are much more opportunistic and willing to try new things.

It's a jungle out there — external economic analysis

Figure 1.5 shows how you need to start your analysis of what is happening outside your firm — in the relatively distant space of broad macroeconomic, technological, sociological and political trends.

Most business plans start (and end) with someone pouring over reams of information about the firm's internal performance over past years. It is important to know your numbers but to break out of old ways of thinking, gain some real insight into the challenges and opportunities that lay ahead you have to look further afield than the four walls of your office.

If you ask any property person whether they know what is going on in the market you can be sure that they will answer a confident "Yes". That's because it is the agent or surveyor's job to know what is going on in the market. The property market in terms of how it affects

Figure 1.5: The onion approach to analysis — start outside and work in

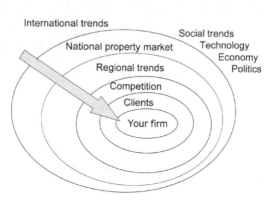

Source: Kim Tasso

house (or commercial building) rents and sales that is — and probably over the short term (up to a year).

But the market analysis we need for a business plan is different in two particular ways. First in terms of the time frame — we are looking at between three and 10 years ahead. And second in terms of the range of topics being looked at.

A helpful exercise is to use some of the younger members of your firm to use the internet to undertake some simple research to identify long-term future trends and predictions (ie, for at least five years ahead) in the following areas:

- Economic
  - What is happening in the global and European markets?
  - What is happening in international trade?
  - Which countries are prospering and which are in decline?
  - What is the outlook for interest rates?
  - How buoyant is inward investment?
  - What is happening to unemployment?
  - How easy are commercial and personal loans?
  - Which industrial sectors are performing well and which are suffering?
  - How are the financial markets holding up?

- Political and legislative
  - What is being discussed in Europe at present? How and when will these things affect the UK?
  - What would happen if there was a change of Government?
  - What new laws and regulations are in the pipeline? Taxes?
  - What is happening with green and environmental policies?
  - What is in the current Government's manifesto that is still to be addressed?
  - How important is regional and local government?
  - What is on the table for imminent legislative change?

- Technological
  - How will automation continue to affect life at home and the office?
  - What is the latest on alternative fuels or vehicles?
  - How will the internet, web and social networking develop?
  - What new medical advances are anticipated?
  - What new electronic media products are anticipated?
  - What is happening in the area of biotechnology?

- Sociological
  - What is happening to the demographics of the country and your region?
  - What is happening to the family? Marriage and cohabitation?
  - What are young people studying at school and university?
  - How will immigration affect social cohesion?
  - How are people spending their leisure time and disposable income?
  - What new forms of leisure activity will become common?

If all this sounds a bit too difficult and far fetched to have an impact on your local property market or your practice, then think again. It was through analyses such as these that the smarter and more forward thinking agents appreciated the impact of the internet and made early moves to establish a market leadership position online and divest of their expensive High Street locations — generating significant income flows from presenting the adverts of others on the internet rather than only selling properties. It was through analyses such as these that some surveyors formed networks giving them access to a larger number of offices and more critical mass to allow the central overheads of sophisticated technology and marketing to come within their grasp. And there are many other examples.

In terms of the research, do not let this be a one-off exercise. Changing your firm (and your people) so that they are constantly scanning the external environment for "weak signals" is a really good discipline to ensure that your firm remains on top of things — and proactively grasping new opportunities rather than reacting despondently to threats. Because if you do not grasp the opportunities early and be one of the bold pioneers then you are consigned to being one of the head-in-sand laggards, who have to grapple with those opportunities when they have become threats to your very existence.

You may have noticed that I have not reproduced an example PEST (political, economic sociological, technological) analysis for you — to show you the sorts of things you should be looking for. Partly this is because such analyses get out of date quite quickly. Partly because what you need to consider may be different to what other firms need to consider. Partly because if I give you an example you will not do the work yourself and your analysis will be flawed. Also because if everyone looks at the same information, everyone will draw the same conclusions and embark on similar strategies.

And where do I get my information? I use a variety of sources. I read *The Economist* and the *Financial Times*. *The Economist* publishes a helpful annual review (The World in 2009), which you can buy around Christmas time that saves lots of internet searching. Then there are many government statistics on the web and specialist sites for particular areas such as marketing, research, technology. The sector/trade press are also an excellent source. As are the better quality blogs (such as those associated with national newspapers). You can even encourage people to read the tabloids as often there will be useful bits and pieces in there too. Really smart firms offer a small token of appreciation to staff who submit cuttings of interesting trends and statistics — and not only does that spread the research and scanning load, but also it gets people involved and it means that your entire workforce is more up to speed on economic, technology and other market trends. If you store them on your intranet, it means everyone will be super-up-to-date on what is happening and the impact on your clients' businesses. How impressed will your clients be at that?

And how do you use this information? You need to think about each selected item of information and consider its impact on your markets and your clients and then what the property implications might be as a result. It is not easy but it is interesting.

Hopefully, your PEST analysis will have opened you and your partners eyes to some of the uncontrollable things that are happening out there that will ultimately have an impact on both commercial and residential property markets over the next three to five years. It may have already alerted you to some opportunities and threats.

Checklist and actions

- Start to review and/or develop a mission or vision statement that all the partners are comfortable with.

- Develop a list of the firm's core values.

- Identify a small group to oversee the business planning process.

- Allocate responsibility for different aspects of the PEST analysis to various people in your firm.

- Set a deadline for compiling the PEST analysis and discussing the results and the likely impact on your business.

- Look at your intranet to see how you can store this PEST information centrally for people to access and update it regularly.

In our neck of the woods — regional analysis

While international and national trends are important, so too are those that are a little closer to home. If your firm is based within one or a number of counties you need to obtain some information about the general economic and demographic trends within your region.

For example, one firm assigned one of its young surveyors to a major piece of research that provided an overview of each of the counties where it had branches. The surveyor contacted various government and business support agencies to gather a huge amount of information about the composition, trends and attributes of each area.

As well as providing a highly informative knowledge base for all members of the firm, showing how different strategies were required in different locations, it also produced a useful by-product in terms of a valuable summary for local people and businesses who were comparing regions — with commercial rents and prices as well as residential rents and prices, growth statistics and employment figures summarised and compared in table 1.1.

Table 1.1: Example regional analyses

	East Midlands	UK
Area (km^2)	15,607	242,514
Population (2003)	4,252,000	59,600,000
Working age population (2003)	2,679,200	36,544,900
Total regional GVA (2003)	£61.7bn	£951.7bn
Mean annual salary	£19,513	£22,248
Employment rate (working age) (%)	75.3	74.1
Unemployment rate (working age) (%)	4.7	5.1
Total number of university students	152,500	2,247,400
Travel to work time (minutes) — 2003	18.35	20.06

Furthermore, as they used some of this information in their PR, marketing and relationship management campaigns the firm became known as an important player in that region and a firm that was interested in local development. As a result, their profile and position amongst the authorities became stronger and they were invited to participate in and then speak at events. This led to some unexpected but welcome work opportunities.

The other thing it showed was how different were the locations of each of its offices and branches. This meant that a "one size fits all" approach — particularly to marketing and client development — was flawed. By looking at the relative growth and attractiveness of the different geographic markets, they were able to get a rational basis for deciding which offices to develop, which to leave alone and which needed more considered treatment.

By looking at each geographic market, they were also able to identify the key players, key competitors, targets, media and other intermediaries and influencers. Again, this built a valuable information and knowledge base against which to take future decisions. Another by-product was that all the partners began to understand the different challenges facing their colleagues in other offices.

There are numerous sources for regional information, but the development agencies are a good starting point:

England's regional development agencies:

- Northwest Regional Development Agency
- Yorkshire Forward
- Onenortheast
- Advantage West Midlands
- East Midlands Development Agency
- East of England Development Agency
- South West of England Development Agency
- London Development Agency
- Southeast England Development Agency

Shops, sheds and semis — sector analysis

While some firms organise themselves geographically, others — particularly those with a strong commercial client bias — look at the business from a property market point of view. For example, where you have lots of work with education authorities and schools, farmers and agricultural businesses, hi-tech and telecoms companies, energy and utility companies or even retailers, it is a good idea to organise some structured research into these sectors.

This helps in a number of ways. First, as a planning tool you start to connect with the commercial issues that your existing or potential clients are grappling with and foresee changes in their requirements or demand trends. This means that you are better able to gear up your resources and position your firm to take advantage of those changes.

Second, it means that everyone becomes more familiar with the business issues affecting companies (developers on the supply side and occupiers on the demand side) so that you come across as more credible professional advisers and also differentiate yourself from your colleagues, who are still talking about property issues as they relate to old market dynamics.

You need someone do this research who is reasonably familiar with the sector — otherwise it is too steep a learning curve. But it is relatively easy to find the websites of the key media and the main trade or professional associations that serve the sector and these sites act as great portals providing the necessary links to the best information sources and market data. The RICS is a good source of information, particularly on rural markets, as is DEFRA. *Estates Gazette* and other property and construction magazines produce surveys of specific markets such as retirement homes, healthcare and nurseries.

Sometimes, you have to invest some money as the best data sources are commercially operated. A good route through this is to ask your clients and contacts in the sector which data sources they rely on and assess your options that way. It may also be worth connecting with an experienced research professional or consultant in the market, who can accelerate your learning curve, show you the best sources of information and maybe introduce you to some of the key players.

Remember at this planning stage that you are gathering information about the external environment and how the trends might impact on your firm and its opportunities in the future. So do not let the research exercise prompt you to make premature decisions and take action too early. If you are working across a large number of sectors, then you will need to make a choice to concentrate on a manageable number. There are some pointers on selecting relevant markets in this section and also in chapter 4 on strategic marketing. So your research work is about finding comparable information and/or information that sets one sector up as requiring particular attention.

At the end of the research exercise, you will need some discussion that summarises the key points and offers some suggestions as to what the main impacts are on your future strategy. And as you probably have not yet developed your new strategy this can be tricky.

It goes without saying that a part of your sector analysis will be a review of the property sector. You will probably have a huge amount of knowledge on this from all your partners. However, the question is whether that knowledge is broad enough and long term enough to support the long-range planning questions that we will be addressing. Often your specialists will know a great deal of detail about very specific issues but little about the big trends and the more uncertain developments that are likely to arise down the line. In this case, it might be worthwhile asking someone who is less familiar with the property sector to undertake a review — as their naïve questions may focus on things that you have taken as given or highlight areas or issues that you have noticed but not really questioned about its future impact.

Sometimes you have to get smart and do your analysis combining a number of elements — for example, comparing the industrial market in the East Midlands with the retail market in the North East.

Do it to them before they do it to you — competitor analysis

In addition you need to carry out a systematic review of your competitors. This means looking, in a creative way, how your residential and commercial clients might address their property advice and/or agency needs in other ways — through the internet, through national firms, through international firms, by doing it themselves, using the property teams in banks, using independents, using accountants and such like.

Markets are constantly changing and there is an increasing number of alternative ways (and often cheaper ways) to do the same things that were once the sole preserve of the chartered surveyor or estate agent. I would advise looking critically at who might be nibbling at the edges of (or taking huge big bites out of) your market.

Some people use a framework called Porter's Five Forces where you look at present and future competition across a number of dimensions. His research into market attractiveness looks at your immediate competitors, but also it considers:

- *Threat of new entrants*
  This includes those from overseas markets who are looking strategically at the UK (perhaps as a stepping stone into Europe), or organisations outside the property sector who look with interest at the senior relationships and/or related investments that your market makes. Often, deregulation and globalisation create changes that allow others to enter your market — despite any protection offered by professional qualifications. Additionally, the internet has seen droves of new and established companies trampling over the property heartland.

- *Buying power of suppliers*
  What do you need to grow your business? You definitely need good quality new recruits and they may be sought by many other players in the market too, thus creating a significant constraint on your growth plans. You might need premises in locations where there is little supply. There may be a lack of good support staff in your area. Also, limitations on your ability to get what you need to provide your service needs to be considered.

- *Buying power of clients*
  Clients are getting smarter — and bigger. They have started to do stuff for themselves. They might even be offering services to parts of your market. They may organise themselves into collectives so that they can buy in bulk, set up restrictive (for the suppliers) framework agreements and force down prices. In-house teams now have more power and influence than ever before.

- *Threat of substitutes*
  Clients may find different ways of meeting their property needs. The internet information explosion means that they are no longer reliant on their property advisers for up-to-date information. They can do things for themselves — particularly as there is so much good software out there. When the markets are generally down, all sorts of other operators may be looking to find ways of persuading clients to spend money with them rather than you. People could hire contract staff or interim staff instead of hiring your firm. They could decide to outsource their property work entirely. You must consider the alternatives that clients have to paying your firm.

Most of your direct competitors will have a website where you can see the range of services they provide and the main markets that they serve. You will be able to see information about their senior people and their backgrounds, age (proximity to retirement) and specialist areas. The news section will also alert you to recent deals and developments. You should also scan the property and local press for mentions about your competitors — and build an intelligence file.

Assign different people to being responsible for watching the activities of nominated competitors so that they become the fount of knowledge for these firms. Most firms will be profiled in the property industry's league tables (surveying firms, rural agents) where you will be able to find less generally available information about their income, profits and growth rates.

Perhaps more valuable than this public information will be the information that you are able to glean from less obvious sources. If you are asked to prepare a tender or a pitch you may learn from the panel discussions (or even the debrief) what other firms have in terms of strategy, key messages and, of course, their work methods, systems and fee rates and pricing policies.

Some firms also organise for new staff who have joined from competing firms to do a debrief to share their insights about the firm's culture or its own perceived strengths and weaknesses and how the competitor firms view your firm. Naturally, you will be alert to the contractual restrictions on your new staff and will not want to put them in a position where they are being asked to divulge information under which there are some confidential or contractual obligations.

Sometimes you will have clients who use a number of different firms and they should be able to provide interesting insights into how they perceive the various players and their strengths and weaknesses. But be careful because if these clients are happy to share their insights about other firms with you, then they are likely to do the same with your competitors about you!

There are a number of reasons why you need information about your competitors. The first is an important marketing idea called "positioning" and "differentiation" — how you will talk about your firm to clients when comparing your firm and its services against theirs and also in terms of recruitment. Your staff may be tempted to move to a competitor that is perceived as being more successful than your firm and similarly your competitors may be the source of your future recruits and/or partners.

When we come to thinking about strategy, you need to consider your position against your competitors. Therefore, it is necessary to understand the main dimensions against which you and your competitors are similar and different. Sometimes it helps to prepare a chart so you can see this spatially — some refer to this as a perceptual map. The diagram below is an example that might help you summarise your analysis.

Figure 1.6: Example perpetual map to help positioning decisions

Source: Kim Tasso

As with the other analyses, depending on the geographic or sector organisation of your firm, you may have to consider different competitors in various markets.

As you conduct this competitor analysis you will also have another agenda. For those firms who are keen to grow by merger or acquisition, you might be thinking which of your competitors might be suitable targets. While the rational data (location, size, partner profits, client base) is important here, so too is less tangible information about the culture of the firm. The culture is not something that is easy to pick up from websites or written reports — you can only glean that information from meeting and speaking to people from the firm. Consequently, that might require you to mix and mingle with them for a more personal and subjective view about whether they are "our kind of people".

Checklist and actions

- Assign people to undertake a systematic analysis of the various regions of interest to your firm — now and in the future.

- Identify and make contact with the local development agencies and other public sector bodies that can provide information.

- Identify the relevant property and industrial sectors for your firm and assign people to undertake structured analyses.

- Design a project to gather information about your present and future competitors.

- Consider how you differentiate and position yourself against your competitors (both now and in the future) — be realistic about the relative strengths and weaknesses.

- Review all the external analyses and identify the main opportunities and threats.

Eagle eye — looking at your business critically

We have spent some time and energy examining the external environment in which your firm operates and also in which it plans to grow in the future. Now it is time to turn your eagle eye onto your own firm and start to undertake a detailed audit of all the relevant internal information.

If your firm is reasonably sized, you will be organised into branches, offices or divisions. It may be that you need to conduct the analysis at that level once you have the broad, firm-wide brushstrokes in place. One way you might achieve this is to ask each office or division to provide information in a similar, structured way by providing a proforma document or setting up a standard series of questions on the intranet that guides them through the process.

A natural starting point is the financial profile. Fees, costs, key clients and profits should be easy to obtain from your accounts system and it is usually possible to gather information for the past three years to provide you with a background of trends. Other information includes the number of transactions or clients, which should then provide an average fees or profit figure that might make it easier to compare across groups. Other areas to consider are the length of the sales pipeline, the average conversion time and rate and the overhead expenditure profile.

The checklist in table 1.2 will help allocate different tasks to different partners. This will spread the load and also make it more likely to gain the interest and commitment of all the partners if they are all involved. Of course you might wish to add other information that is relevant for your firm and there may be some areas where you do not have the required information easily to hand. Do not worry about exactly what information you do and do not have — it is rare that the firms I deal with are able to easily produce all the information that I request. However, you might note down where there are serious gaps, as plans to fill these gaps might become part of your future strategy and development plans.

A picture is worth a thousand words

You probably now have a huge amount of information in a number of files. The trick is to review and summarise the information to make the insights it provides apparent. There is no substitute for some charts at

Table 1.2: Example internal audit checklist

Type of information required	Specific information sought and questions to ask	Key observations or lessons learned or issues to discuss
Plans	❑ Previous and current business plans ❑ Current year financial projections ❑ Staff, IT and marketing plans ❑ Material from recent partnership meetings ❑ Details of past or current merger discussions	
Firm structure	❑ Organisation charts ❑ List of offices or branches with a brief profile (eg, range of services provided, fees, number of staff, location) ❑ List of divisions with a brief profile	
Partner information	❑ Partnership deed and structure ❑ Management or partnership meeting minutes ❑ Number of equity, salaried and other partners ❑ Remuneration structure and systems ❑ Individual partner profiles (and plans if they are available) ❑ Leverage — ratio of partners to fee-earning staff ❑ Age profile	
Professional staff information	❑ List of staff by department, office and specialism ❑ Fee income breakdown by staff member over past few years ❑ Seniority profile (identify any gaps in post-qualification experience bands) ❑ Ratio of fee-earning to secretarial/support staff ❑ Breakdown of staff by department ❑ Future partner prospects (succession plans) ❑ Recruitment plans ❑ Gender and race split	
Finance	❑ Fee income and gross profit by division and/or office and/or service ❑ Work in progress or deals pipeline ❑ Credit control procedures and systems ❑ Current bank balance, loans and cash position ❑ List of aged debtors ❑ List of key clients (top 100 if commercial) and income for past three years ❑ Budget and actual for past three years ❑ Risk assessment policies and procedures	
Technology	❑ Main technology platforms ❑ Key systems in use — and performance comments ❑ Annual expenditure in systems, maintenance and training ❑ Knowledge systems, intranet and extranet ❑ Own systems that were developed for particular applications ❑ Profiles of the IT staff (or details of outsource agency) ❑ Levels of technology awareness and use throughout the firm ❑ Planned upgrades and developments	
Marketing	❑ Past and current marketing plans ❑ Corporate identity guidelines ❑ Past and present marketing budget and expenditure ❑ Structure of marketing department and profiles of staff ❑ Review of external specialists used (and contracts and expenditure history) ❑ Key brochures, newsletters and advertisements ❑ Schedule of events ❑ Details of the marketing database	

this stage, as it makes it easier to assimilate the broad trends and to view comparisons between various elements of the business. Here is a reminder of the sorts of charts that you may find helpful:

Figure 1.7 Example fees and profits — five-year trend (£m)

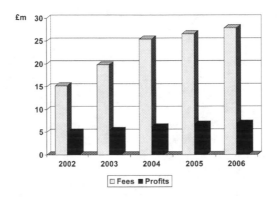

Source: Kim Tasso

Figure 1.8: Example fees — regional breakdown, 2006 (£)

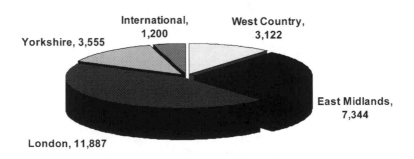

Source: Kim Tasso

Checklist and actions

- Speak to your support teams — accounts, marketing, technology and HR — and explain what analyses you require and when and why. Involve them in the business planning process.

- Ask each office or team to complete a proforma analysis so that you can compare them.

- Consider what information you have about the various sales pipelines.

- Use the checklist and assign individuals to collecting the internal information needed.

- Get some help in producing charts to help everyone understand the numerical analyses and to see key trends more easily.

- Summarise all the analyses and identify the relevant strengths and weaknesses.

- Arrange some time for the partnership to become familiar with the key findings and to discuss likely strategies.

Root and branch — drilling down to the detail

Having looked from the viewpoint of a helicopter at your firm, you now need to scale down and start looking at the critical component parts. This is where you need to start summarising and comparing — below is a simple starting point.

Table 1.3: Example comparison of branch offices

Office	Res?	Com?	Other?	Fee income (£)	Staff	Fees per staff (£)	Comments
Location A	Y	Y	Y	251,444	6	41,907	
Location B		Y	Y	1,099,888	23	47,821	High staff turnover
Location C			Y	645,878	15	43,058	Fastest growing
Location D	Y	Y	Y	1,887,654	17	111,038	Recent merger
Location E	Y			300,065	3	100,021	
Total				4,184,929	64	65,389	

A further benefit of the strategic planning exercise will be that everyone becomes more familiar with what their clients and colleagues in other departments do and therefore further collaboration and cross-selling is promoted.

Profit

As I explained in the introduction, this book omits three important parts of management — a) the operational management of the surveying, agency and other professional work, b) the development and implementation of an IT strategy, and c) detailed financial planning and accounting.

It would be terrible if I were to cover the topic of business planning without at least mentioning profit — as this is the main driver of most partnerships. Subsequently, I have included one critical idea for you to think about profit:

Figure 1.9: Profitability formula for a professional firm

$$\frac{\text{Partners}}{\text{Profits}} = \frac{\text{Profits}}{\text{Fees}} \times \frac{\text{Fees}}{\text{Staff}} \times \frac{\text{Staff}}{\text{Partners}}$$

Productivity

$$\text{Productivity} = \frac{\text{Fees}}{\text{Staff}} = \frac{\text{Fees}}{\text{Hours}} \times \frac{\text{Hours}}{\text{Staff}}$$

$$= \text{Value} \times \text{utilisation}$$

Source: David Maister

Walking the talk — ask the partners what *they* think

While you are assembling and analysing all of the above information, it is helpful to do some more informal research among the partners (and senior fee-earning staff and possibly a representative selection of even the more junior staff and secretaries). After all, as you will see in the next section on leadership, a leader develops a vision only after receiving input from his or her partners.

As a consultant, one of the first things I do is arrange to have a one-to-one meeting with each of the partners for about an hour. Partly this is to get to know the firm I am working with better, but mostly this is because I glean a great deal of information that is relevant to the business planning process by doing this.

I usually send a structured list of topics to each of the partners before I speak to them so that they have time to prepare their thoughts. These topics usually start with their views of the big picture (the market overall), then the firm, then their office and/or division and finally their own practice area. I ask people both about what is currently happening as well as what they think should to happen in the future. Of course, some people prefer to start the discussion at the bottom level (their own practice) and work up!

I keep the rules of this interviewing process fairly tight. While I write down everything that everyone says and add it to the growing pile of information that I have gathered as part of the research and planning process, I prepare a transcript that I then offer to them to review and change as necessary. This gives them the opportunity to add or amend their comments, and to remove those that they do not want publicly attributed to them. It also means that when I come to present my findings, a pile of approved transcripts is available for all partners to look at and compare with their own views. Often this in itself is a hugely valuable process promoting greater understanding and knowledge between partners and supporting the start of better internal communications and even, wait for it, a first step towards collaborative working and the elusive holy grail of cross-selling.

Sometimes, you can ask partners to think about setting some preliminary goals during the interview process — although caveat this that you may not accept their first offer!

Once all the "interviews" are completed, it is time to lock yourself into a room with a towel around your head while you attempt to tease out the main trends, the areas to investigate further and compile various lists such as: key findings, areas where everyone agrees, areas where there is a difference of opinion, further areas to investigate and such like.

Table 1.4: Illustration of linking firm-wide strategy to teams and partners

Mission	Objective	To be achieved by	Main actions needed to meet this goal	Comments
For the firm overall				
For my office(s) or division				
For me personally				
For my team members				

SWOT — Strengths, Weaknesses, Opportunities, Threats

Summarise all your analyses with a SWOT. If you have followed the above instructions correctly, then you will find yourself awash with information. So how do you sort the wheat from the chaff? While having plenty of information is important, you also need to identify the dozen or so critical elements that will affect your firm — positively or negatively — that you absolutely have to address.

A tool that might help here is the SWOT — Strengths, Weaknesses, Opportunities and Threats. Figure 1.10 shows what to do.

Figure 1.10: Summarising your analyses — SWOT

Internal	Strengths • Point 1 • Point 2	Weaknesses • Point 1 • Point 2
External	Opportunities • Point 1 • Point 2	Threats • Point 1 • Point 2

A dirty weekend

How did I know that you would turn first to this section? Go back and read from the start of the chapter!

Once you have gathered all the relevant information together and a small team has spent some time pouring over the figures and trying to get to grips with it all by producing a SWOT, a key driver sheet or other summaries, I suggest that you get the entire partnership together for a weekend in a country house hotel (or something similar) so that everyone can examine the data and contribute their ideas and thoughts for inclusion in the plan.

The sort of agenda you might have to structure the discussion around could look like this:

1. Introductions and aims.

2. The business planning process.

3. Review (reminder) of the analysis:
    *   firm-wide;
    *   financial;
    *   office or division comparisons;
    *   external analysis key findings.

4. Preparing the business plan:
    *   mission and objectives;
    *   key challenges for:
        –   finance;

- human resources;
- marketing and new business;
- client development;
- information technology;
- premises and facilities.
  - Options and decisions.

5. Moving forward:
   - action plan — who does what and by when;
   - monitoring progress;
   - next steps.

It will make it much easier for everyone concerned if you have one or two people (an external person is a good idea — provided they understand the nature of your business, are sensitive to the needs of all the different partners and excellent as a strong but inclusive facilitator) who can present the information in a sensible order, highlight the key points, guide everyone to look at the information and debate the key issues and ensure that some decisions are reached.

If you have a large partnership, it is a good idea to organise some time at the weekend retreat in groups so that people get to know one another better. While there are merits in having break out groups for particular offices and divisions, there are also advantages in mixing people up so that each group has a variety of disciplines, offices and age profiles represented. If group work is undertaken, you must ensure that there is suitable report back time that allows the groups' thoughts to be debated by the entire partnership.

The advantage of preparing the business plan this way is that everyone is familiar with the information and the process and has had a chance to make a contribution and understand the rationale for various decisions. If people feel they have been involved in the plan, then they will be more likely to work to implement the agreed actions.

Of course, you also need someone to be ready to write down all the points that are raised and discussed. If you use flip charts and handouts during the weekend then this will be made easier. Make sure you allocate this task before you start or you will "lose" lots of valuable insights. Too many property partnerships have great discussions but fail to document the actual decisions as to who is going to do what and by when.

Aim for the sky. But avoid the middle ground

One of the most challenging things when you come to analyse all the data and review your mission and overall aims for the firm is the way in which you set some SMART (Specific Measurable Achievable Realistic Time-specific) goals.

Most firms will find it easy to set headline financial goals — how many fees and how much profit per partner. But there are other objectives here too — the shape of the business in the future, and this will indicate that you are aiming for higher growth in some parts of the firm than others.

Human resources goals are relatively straightforward too — how many qualified staff, with what speciality expertise and how many secretarial and support staff you will need to help them. In many respects, the people targets will be based on the financial goals.

Then the objectives become a little trickier. For example, how do you want your client mix to change? Some clients may generate lots of fees but little profits. It may be that one-off transactional work generates a better profit but with a much greater risk. There may be regular, profitable work that no one enjoys doing. You may need objectives around work or client mix to guide your strategy decisions.

Figure 1.11: Forecasting and budgeting

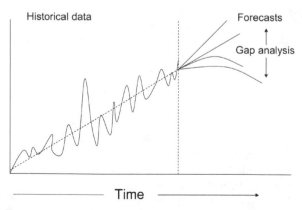

Source: Kim Tasso

Figure 1.12: Setting SMART objectives

Objective	How to measure?	Date to complete?	How to monitor?	Realistic?	How?

Source: Kim Tasso

Another area is that sometimes people become confused between aims and strategies. An aim is something you plan to achieve, whereas the strategy is how you plan to achieve it. Hence, your aim might be to improve profitability by 15% and the strategy will be to focus on your larger existing clients and only a handful of major prospects. Alternatively, your aim might be to increase the number of new enquiries generated by 30% — the strategy to do this will be within your marketing strategy (see chapter 4).

Avoid the middle ground — shrink or grow

If you are neither a large practice, nor a small niche practice then you are in danger of being in that challenging place called "the middle ground". Here you have neither the economies of scale or critical mass to compete on price or with the resources of your larger competitors, nor the specialist expertise to support premium prices.

In the middle ground you have to make tough decisions. It may be to reduce the range of services you provide, or close down some of the least profitable or stagnant offices. You may have to look at changing the nature and shape of your workforce — either through expensive training programmes or lateral hires.

Figure 1.14 may help you identify the appropriate strategic thrust for your practice.

Figure 1.13: Mid-sized firms suffer lower profitability

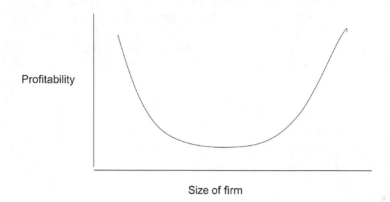

Source: Kim Tasso

Figure 1.14: Ladder of growth

9. Growth outside
industry boundaries

8. Acquisitions and alliances

7. International growth

6. New distribution channels

5. Enter new markets

4. Develop new products and services

3. Win new clients

2. Grow share of client

1. Increase client retention

But you should ensure that your strategy does not focus on just one area to the exclusion of others, since whatever you do will have an impact on other areas. Some firms have found it useful to frame their business plan in terms of the goals and measures on the balanced scorecard, which was developed by Kaplan and Norton. It has the advantage of recognising the links between traditional functions, such as finance, technology and marketing.

- Financial perspective
    - What is important to the partners and staff?
    - Revenue growth?
    - Profit growth?
    - Cost management?
    - Working capital and cash flow?
    - Asset utilisation?

- Internal business perspective
    - What must we excel at in terms of our operations?
    - Where are our greatest efficiencies and inefficiencies?
    - What do we buy-in and what we do ourselves?
    - How can we provide our service(s) better?
    - Which processes are critical?
    - How do we manage the work? Relationships?

- Innovation and learning perspective
    - How do we continue to improve and create value?
    - What are our employees' capabilities?
    - What new knowledge and skills do we need?
    - Are we deploying technology effectively?
    - How do we maintain motivation and commitment?

- Client perspective
    - How do we want clients to see us?
    - Which are the most profitable clients?
    - What is our market share?
    - Which are the critical client (and intermediary) relationships?
    - How do we assess client satisfaction?
    - How do we acquire the right sort of new clients?
    - How do we retain and develop existing clients?

One of your strategies might be to look for a marriage partner with whom to merge or some smaller fish that you might gobble up. Mergers are covered further in chapter 9 on change management. There are a number of other strategic approaches — mainly concerned with the mix of markets and services that you are targeting (see chapter 4 on strategic marketing).

Diversity, social responsibility and all that jazz

The younger generation (your potential future workforce) and increasingly clients will be interested to hear what you think about beyond the pound signs. There are also more regulations around having a diverse workforce. Some clients (particularly in the public sector) take these things very seriously. Larger corporations and your local community will also be interested in your views towards social responsibility and issues such as the environment, sustainability, waste management and recycling policies. Indirectly, these things will impact your strategies and the resources you have available to help you grow and perhaps even in the rate at which you can grow or increase profits.

Some firms may place these issues at the heart of their business plans, while others may only do the bare minimum required by law. It does not matter as long as you have considered their importance to your present partners, your future partners and your present and future clients.

Checklist and actions

- Arrange a one-to-one interview with each partner to talk about the business planning process and to specifically illicit their firm, team and personal goals and key thoughts about the future of the firm. Document these discussions and allow them to be edited/approved for circulation to other partners.

- With your business planning team, spend a day or two away from the day-to-day demands of your job going through all the external and internal analyses and prepare either a SWOT analysis or a list of key drivers of the business.

- Circulate a summary of the various analyses — preferably with a SWOT analysis to focus on the key issues and an agenda for a partner awayday (weekend).

- Schedule and prepare a weekend when the entire partnership can talk through the results of all the analyses and the preliminary recommendations for the business plan.

- Review your forecasts and budgets in the light of the various analyses and make adjustments as required.

- Review your attitudes and policies with regards to diversity and social responsibility.

- Set some SMART goals for all aspects of the business.

Selecting the appropriate strategies

The appropriate strategy will depend on the results of your external and internal analysis as well as on your goals. However, there are some generic strategies that may depend on the particular phase of development of your firm. Greiner had some useful insights. He mapped out the phases of growth and identified key evolutionary and revolutionary phases depending on an organisation's age and size. The various phases — and appropriate strategies — he suggested were as follows:

Table 1.5: Phases of growth and strategies

Phase	Growth strategy	Crisis
1	Creativity	Leadership
2	Direction	Autonomy
3	Delegation	Control
4	Collaboration	Staff
5	Co-ordination	Unknown/Various

Source: Greiner

The rest of this book should help you to formulate the various top-level and detailed strategies that you wish to pursue in order to help your partnership grow and flourish. These strategies might be corporate — merger or acquisition, marketing — developing new markets or services, promotional or through the development of your people. The remainder of this book offers some thoughts and guidance in these areas. Once you have selected appropriate strategies you will need help to implement them — so the chapters on leadership (chapter 3) and change management (chapter 9) should provide some food for thought there.

It will be easier to identify what the appropriate strategies are likely to be if you have done good foundation work with the analyses outlined in this section and set some specific aims and objectives.

On the brink of destruction — turnaround strategies

But what about the recession and those firms that are now pushed to the brink of destruction? Some of the ideas in this book may help but sometimes more urgent or dramatic action is required if you are to avoid calling in the liquidators. Naturally, you will seek the advice of your usual accountants — and the bank. But when we look to the management experts we can consider the hard steps Slatter suggested for "turnaround" situations:

- *Gain full control of the situation* — This includes communication with the key stakeholders (including the bank) to establish your credibility and to reassure them that you are aware of the problem(s) and are dealing with them. The worst thing that you can do is to stay in denial, avoid tackling the problem or hoping that things will simply improve on their own. You need to step away from your usual day-to-day activities and take control of the situation.

- *Assess the situation* — Face the facts and take action. Assess (and replace if necessary) the present management and advisers. If you need additional external expert help, get them in.

- *Evaluate the business* — The business must be fully evaluated. The first pass means applying the Pareto principle of finding the 20%

of the factors that are having 80% of the impact to allow management to focus effort, develop appropriate action plans and implement organisational change immediately. Identifying the most profitable clients, services and teams is critical.

- *Motivate remaining staff* — The staff that have been retained need to be reassured and motivated. Regular communication is important.

- *Install budgeting and control systems* — Establish central weekly cash flow information and ensure that the most senior partner signs off all capital and revenue expenditure.

- *Consider next steps* — Consider other strategies, such as merger or diversification.

To support this, Grinyer's "Sharpbenders" analysis showed that the most common actions by those most successful in turnaround situations included: major changes in management, stronger financial controls, new product-market focus, improved marketing, intense effort to reduce costs and windfalls.

Structure follows strategy

In chapter 3, we look at leadership and management. And a key question is how you structure your firm as a result of your new business plan and strategy. Your structure must follow your strategy. If you choose a strategy with a geographic focus — then your structure must be geographic. Whereas if your strategy focuses on particular markets or service lines, then that should form the basis of your divisional structure. Sometimes there are complex strategies that require what are known as matrix structures where, for example, you might be organised both by geographic area and also by specialist service type, such as valuations, building consultancy, office agency and such like.

There are many different ways to organise your practice, consider the following alternatives:

- *Centralised versus decentralised*
  Do you want to maintain all decision-making and systems in the centre, at head office, or is your firm large enough to need management teams and decision making at the divisional or office level?

- *Hierarchical versus flat*
  Larger organisations have numerous levels of management — with juniors reporting to associates, associates reporting to partners, partners reporting to divisional heads, divisional heads reporting to regional leaders and finally regional leaders reporting to the Board. Or is your organisation fairly flat, with all partners reporting in to a central team or Board?

- *Functional versus geographic*
  As mentioned above, some firms organise themselves with divisions focused on particular areas (and these are likely to be multi-disciplinary teams) while others will have departments of the same types of specialists.

- *Command and control versus organised around goals*
  Do you want to tell everyone what to do in a high degree of detail or are you more comfortable with setting out the broad aims and allowing the divisions or teams to decide how best to achieve them?

- *Authority based versus driven by values and beliefs*
  Do you need to tell everyone what to do and how to do it or does your firm have sufficiently strong values and beliefs that people know the way that things should be tackled and can be trusted to do so on their own initiative?

- *Collective responsibility versus individual accountability*
  Some firms organise themselves as teams and assess and measure team performance whereas others will focus on the contribution of each individual. Typically, smaller firms focus on the individual accountability of partners whereas larger firms focus on the performance of teams.

- *Single line reporting versus matrix reporting*
  This was mentioned above. You may just want people to report into a divisional structure and monitor progress on that basis, or you may require people to consider their performance in a number of ways — for their region, for their markets or for their particular services. It may be that people have to report into a number of different partners in each of the different areas.

You will need to match the structure to the style of the organisation as well, consider which of the following is most like your firm — and perhaps look at some of the material by management experts, such as Charles Handy, to gain an understanding of the particular features of these organisational structures and how best to organise and manage them effectively:

- Corporate style
- Entrepreneurial style
- Bureaucratic style
- Family business
- Small business
- Public service/armed forces

The way in which you structure your firm is considered further in chapter 3 on leadership and management.

Using your business plan

Once you have completed the first draft of your business plan, you will naturally wish to share it with the other partners to obtain their comments, ideas and suggestions. You might also use the exercise to identify "volunteers", committees or project teams who will take the lead in implementing the various programmes and actions.

Once the business plan is finalised, you need to also ensure that the key elements are communicated to all members of the firm — perhaps at a staff meeting or alternatively with divisional or office managers leading more interactive discussions at their regular team meetings.

The Board must ensure that it schedules time to review progress on the business plan on a regular basis. Sometimes this will require further information to be obtained (and this can involve having to set up new information systems to collect and report on the information that you are monitoring as a result of the new goals and strategies) so attention must be devoted to this.

Some firms like to schedule separate (quarterly) meetings specifically with the intention of hearing about how implementation is progressing — perhaps with special reports from the various partners and teams leading the various initiatives.

Other firms — particularly those who mix billing methods, such as transactional percentage based methods with hour or daily fee rates, might find it useful to summarise the position for day-to-day management with a "key drivers" sheet. A one-page summary that provides all the information that they need in order to see — at a glance — how everything is progressing and where the deviations from plan are occurring. This makes it easy to review quickly and easily how everything is going — and where problems might be emerging.

What I am trying to say here is that only once you have completed your plan does the work really start. The plan is a living tool and you must continue to review and refine it as time progresses.

Checklist and actions

- Once you have identified your main aims and objectives, consider all the various options available to you to achieve them.

- Document all your analyses and decisions in an outline plan and circulate for all partners to review and discuss.

- Once you have agreement, communicate the main elements to all your staff — and invite their comments and contributions.

- Allocate regular time at your board or management meetings to review progress against the plan.

- Develop a structure that follows your strategy and the style of your firm.

Case study: Stiles Harold Williams

Stiles Harold Williams (SHW) is a general practice surveying business with seven offices in London, Surrey and Sussex. Fees of £11.3m were generated in the year to 30 April 2008. David Hadden has been with the company for 20 years, having been appointed head of the property management division in 1991 and managing director in 2000.

History
"In 1997, we had been owned by Alliance+Leicester (A+L) Building Society for almost 10 years and were clawing our way out of the last recessionary period, making fees of less than £4m a year and losing money. The society had

parachuted in a managing director who had previously been head of a well-known fast-food chain in Europe and North America, who knew little about property and less about the UK market. However, he taught me about focusing on the bottom line — the trouble was that, for SHW at that time, the line was usually red! He also taught me that the only reason we come to work each day is for the benefit of the shareholders. However, as time progressed and things were not working out too well, when it was announced that A+L would demutualise, we knew SHW would not be core to the new business and a group of 15 directors, myself included, saw our chance to effect an MBO (Management Buy Out).

One of these directors, Robert Stiles, took us through the MBO and agreed to the role of chairman and chief executive, but for a three-year term only. We did not have much of a plan for the future and most of us saw the MBO as a way of protecting our jobs. However, we knew that we did not want to be run by someone who had not always been at the heart of the property business and was not "one of our own". After three years at the helm, in 2000, Robert moved into the role of chairman and I was elected managing director.

In those early days, our aims were modest — we did not intend to double in size or anything drastic — just to grow the turnover and return to profit with steady, organic growth, adding on to the business at the margins and cross-servicing our clients' needs. That is still part of our psyche today — although we needed an agreed plan to focus our attention on the right type of strategic development and direction.

Over the past 10 years since the MBO, we have achieved a steady improvement in both income and profit, we have re-opened offices, where previously A+L had closed them, and have been welcomed back by the business community. Our task is now to maintain our margins and market position while the present uncertain economic and challenging market conditions of late 2008 prevail.

Start with a corporate plan

Our corporate plan is a relatively simple document, containing the broad strategic aims of the business. There are quantitative and qualitative benchmarks to inform us how we are doing along the way; each plan lasts for three years and the most recent one was produced in 2006. We had our own PIG (Performance Improvement Group), made up of key fee earners, who looked at where we were and what we wanted to achieve; we looked at what our competitors were doing and started the plan-making process with a SWOT analysis.

At the time, the agency side of the business was riding high, but we were concerned that things might change — some had already called the top of the market and so a key part of our strategy was to shift the emphasis away from transactional work, towards more generic, professional services work. This did not mean shrinking the agency side of the business; rather, we wanted to grow the professional services side of things at a faster pace. We looked to firms like

GVA Grimley and Drivers Jonas, much of whose performance was driven by professional, non-transactional work. They were more what we aspired to in our own region and seemed more akin to SHW than firms such as JLL and Savills, with their mouth-watering investment deals driving huge profits.

SHW was achieving many of its plan objectives and the second plan period was under way when, in 2004, two major developments unfolded; first, our office agency director in London moved to one of the national agency firms; then, not many months after, our two investment agents in the London office left to set up on their own. With the market so active, we knew it would be expensive to replace them and impossible to attract new recruits into the business, as all good agency and investment brokers were either ensconced in the big or niche firms, or working for themselves. These directors knew our strategy was to focus the emphasis of our business on non-transactional work and, quite understandably, they took advantage of an extremely buoyant market to make their moves; maybe our strategy accelerated their departure, who knows?

During this period, in support of the strategy, our property management and building consultancy teams were beginning to successfully feed off one another. London and the regions were winning good quality management instructions and these were spinning off valuable building surveying work. Our strategy was working for us, allowing us to commit resources to grow the professional side of the business. We deliberately never did try and replace the lost agency work in London and this decision was supported by the plan we had in place. The strategy was working for us.

A few years ago, the balance in SHW was 60:40 transactional to non-transactional fee work; last year, to April 2008, we had shifted the emphasis to a ratio of 30:70, which is the right balance for us in these uncertain economic times. We may have a different view of what is the right ratio in 12-24 months' time, if the market is in recovery, but we shall always maintain a strong emphasis on the professional side of our business, while not losing our market share of agency work in the core territories.

Strategic location and role of offices

We have always known our USP (unique selling proposition) is regional dominance in the South East, mainly along the A23/M23 corridor, and we jealously guard this. Our strategy has been to strengthen the territorial boundaries of our coverage and to expand into adjacent territory by means of small acquisitions and mergers. This strategy also serves to protect the core business with outlying offices. So, for example, we have Worthing and Eastbourne feeding off and into our Brighton office at the centre. These smaller, outlying offices act as barriers to the competition, though they must be profitable in their own right to survive. We have a similar situation in Croydon, with Epsom to its west, although we do not currently have an outlying office east of Croydon. This is a work in progress that we are considering due to the increasing work flow that comes to us from Kent.

Building on our core services with sector and other specialisms
SHW's five core service offerings to its clients are:

1) commercial agency and investment;
2) professional services, which includes Landlord & Tenant, rent reviews, valuations and rating;
3) property asset management;
4) development consultancy and town planning;
5) building consultancy.

One of our office locations, close to Gatwick Airport, has allowed us to develop an aviation specialism and we now undertake both agency and professional work at most of the UK's airports, for major carriers and airline providers. We also have a health sector team, who advise the lending institutions on developing doctors' surgeries, providing valuations and acting for both landlords and tenants on their rent reviews. Our strategy of offering value-added services to existing clients from existing resources was evidenced when one of our valuation surveyors used his contacts with a major lending institution, which also lent into the healthcare market, to secure this specialised valuation work. As a result, we decided it was worth focusing resources full time on developing the work.

I do not like departments of just one person and so we soon supplemented a single resource with a supporting colleague. Similarly, we have a charity sector team, which came into being when we were appointed by the Charity Commission as an approved intermediary for its receivers and managers. We now boast a long list of charity sector and not for profit client organisations.

In June 2007, we recruited our first Chartered Town Planner, having previously enjoyed a fee-sharing arrangement with a planning practice. Our new planner, who had previously worked at senior levels in the public sector, has generated from an almost cold start enough profitable fee work to enable us to recruit a senior planner into this team, which sits within the Development Consultancy Division.

In conclusion, over the years since the MBO, we have learned that it is the relationships nurtured with our loyal clients and our understanding of how their property supports their own business needs, which has brought both the greatest benefit to our business and satisfaction to our people. We use the strapline of "Making property work" to convey the message that we know what is important to these client organisations and the individuals whom we advise.

We have learned that planning for the future is necessary and that the plan has to be backed by the fee earners at the coal face who bring in our profits and those who support them. It seems we can never communicate enough internally, though I often wonder why people seem to want to know everything, only to then ignore what has been said! Communicating with clients, by keeping them informed of what you are doing, even if not much is happening, is vital and never more so than when there are fewer active clients in a slackening market.

For my part, I was not trained to manage the business and, although I have had great support from two successive Chairmen, there have been times when I have definitely flown by the seat of my pants! I have always insisted on retaining a certain amount of my own fee work — I manage a number of property portfolios, mostly for private investment companies and charities or trusts — because I do not expect to be MD when I eventually retire. I try to keep a gentle hand on the tiller of the business, and I try to extol the virtues of inverted management, that is to say, I tell junior staff to look after the relationship with their line manager, rather the manager having to manage the individual. I believe this creates a better motivated workforce and a good atmosphere in the workplace.

SHW can trace its roots back more than 200 years and I like to think we have a long and prosperous future ahead. Mr Blake started his auctioneering business in South Croydon in 1798 and I will not be surprised if, in as many years again, we are still "making property work" for our clients in London and the South East, although I won't be around, of course!"

David Hadden
November 2008

Case study: Douglas & Gordon

Douglas & Gordon Limited (D&G) is a full service estate agency serving Central and South-West London. Its activities include residential property sales and lettings, as well as lettings management, block management, formal valuations and residential property refurbishment. Formed in 1958, it now operates from 15 offices with about 170 staff.

Key performance areas
D&G views the training, development and morale of its people as the most important element in its "armoury", while also placing high importance on innovative marketing and the way its offices are presented to the public, having nearly completed a three-year refurbishment of all its sales and lettings offices costing around £3.2m so far.

Staff recruitment, training and development are taken very seriously and D&G believes it is probably unique in maintaining a policy of the chief executive and managing director personally carrying out a performance appraisal interview with almost every member of staff every six months.

Compiling the mission, vision and values statement
A mission statement was introduced some years ago, but in summer 2008 the D&G Board of Directors decided that it would be beneficial to ask a staff committee to review that statement and also suggest what the company's vision and values should be. The board recognised that an important benefit of a values statement

is to help people understand how they should behave and that the process of drawing up the statement was nearly as important as the end result. Some companies simply generate a statement at board level and then email it to the staff.

In D&G's case, all staff members were sent an email seeking volunteers from each department and level of seniority to form a staff committee for the purpose of reviewing the mission statement and proposing the vision and values, with the committee being chaired by a director. Sufficient volunteers were found to form a committee of 10 people, plus the chairman. They were sent a briefing paper explaining in some detail what they were being asked to do, together with examples from other companies, taken from the internet, and a meeting was arranged.

The original mission statement was:

"Our goal is to reach and retain the position of market leader in every sector of our business by combining uncompromising personal service with expert knowledge."

After detailed discussion, the committee felt that, while that statement was good it should be updated since it would now be backed up by the vision and values. The committee considered all aspects of how it was felt management and staff should behave and its proposals were sent to the Board, who suggested some minor amendments, which the committee agreed, the resulting statement being:

Our mission
To be the most trusted and successful estate agency in London, never compromising our personal and outstanding service.

Our vision
Always being a company people are proud to work for and proud to be associated with.

Our values
People
We encourage each employee to develop to their full potential and to achieve their maximum personal contribution.
We provide career development opportunities and reward employees who show initiative and high performance.
We provide inspirational leadership.
We treat every person with care, honesty, respect and understanding.
We treat every person as we would wish to be treated.
We deal with people in a clear and straightforward way.

Reputation
We manage all our resources in such a way as to enhance our reputation with employees, customers, suppliers and the community.

We provide an environment that encourages loyalty from customers and employees.
We project a contemporary image with traditional values.
We redefine people's expectations of dealing with an estate agent.

Operational excellence
We conduct our business with sincerity, professionalism and expert knowledge.
We honour our promises and commitments.
We conduct our business in an ethical manner.
We work hard to provide an uncompromising service.
We do that little bit extra to differentiate ourselves from our competitors.
We ensure the highest quality of presentation both by employees and in our documents and marketing.
We achieve the best possible results for our customers.

Focus and communication
We listen, understand and anticipate the needs of our customers.
We communicate regularly with our customers.
We are open, approachable and transparent.

Innovation
We are constantly innovative.
We set the standard for our industry.
We are leaders rather than followers.
We adapt and evolve for the benefit of our customers and employees.

Teamwork
We work together across all departments to provide a co-ordinated service to our customers and each other.
We provide a fun, happy and supportive working environment that brings out the best in every team and individual.

Obtaining staff "buy in" and understanding
Together with assistance from committee members, the relevant directors then held meetings with their teams to go through the statement with them to explain the thinking behind each item and seek to obtain "buy in". The feedback was very positive.

Conclusion
The first letter of each value's heading spells "profit", since it is recognised that profit is the company's reward for its people behaving in the right way. It is too early to say how beneficial the above will prove, but D&G feels that the process of involving staff at all stages of the development of the statement has already been very useful.

Leading the way and managing action (leadership strategies)

Roadmap
- Who's the boss?
- To lead or not to lead? How is the question ...
- The helicopter view — strategic thinking
- Walk like a man, talk like a man — developing confidence, credibility and communication
- Where do I start?
- Wanna be in my gang? Picking a management team
- Meetings, bloody meetings
- Power, persuasion and influence
- Delegate don't abdicate
- Amaze as an ambassador
- Bring in the experts
- Banging heads together — managing conflict
- Mr (or Mrs) Fixit — creative thinking and problem solving
- Don't turn a crisis into a catastrophe

"Being powerful is like being a lady. If you have to tell people you are, you aren't."
Margaret Thatcher

"The best leaders are the best notetakers, best askers and best learners. They are shameless thieves."
Tom Peters

Who's the boss?

If you are reading this book, I guess that it is either you or will shortly be you. Congratulations! But being the boss has its challenges as well as its rewards. When things are going well everyone else takes the credit but when things get a bit tricky then it is the boss's fault.

Most property businesses are partnerships — where everyone is equal. And this makes the leadership challenge so much more difficult — and yet rewarding. Maybe you have a managing committee or a management board where several partners share the management load. But someone, somewhere will be the boss — whether this is the senior partner, the managing partner or, if you are very modern, the chief executive.

But this section is about leadership. A firm without a leader is just a group of individuals drifting along. These days, there are so many threats, changing markets, aggressive competitors, new technologies and demanding clients that there is no room for complacency and an even greater need for good leadership.

But leadership, as with so many things in business — is an elusive concept. We all know what good (and bad) leadership looks like but will find it hard to describe. It is often easier to talk about leadership in terms of analogies. I particularly like those concerned with sailing. For example, "Leadership is about capturing the imagination and enthusiasm of your people with clearly defined goals that cut through the fog like a beacon in the night" and the more ominous: "Anyone can hold the helm when the sea is calm" or even, from Faye Wattleton: "The only safe ship in a storm is leadership".

Effective leaders come in many different shapes and sizes. I have met many great leaders during the years that I have worked in the property business. Perhaps the best was a leadership team where the senior partner had the traditional leadership skills — gregarious, respected, liked, visionary and motivational (a people person) and the managing partner who has highly organised, thorough, analytical, cautious and methodical (a systems person).

To lead or not to lead?
How is the question ...

What is leadership? Often, it is confused with management. But management is about planning, organising, leading and controlling, as shown in figure 3.1.

Figure 3.1: The elements of management

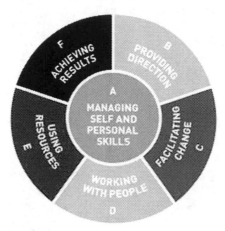

Source: *www.management-standards.org.uk*

For my friends in the rural sector, the word "management" comes from the French *manège* (Italian *maneggio*), which means dressage exercises to train horses in obedience and deportment! Early models confused management and leadership but saw the main dimensions as the focus on people and the focus on getting the job done.

Table 3.1: Leadership styles

Country Club management
High concern for people
Low concern for objectives

Team management
High concern for people
High concern for production

Impoverished management
Low concern for people
Low concern for production

Authority compliance management
Low concern for people
High concern for production

Source: Blake and Mouton

Over the years, there have been many theories about leadership. Originally, there was a view that a leader was a composite of various personality traits such as: drive to achieve, motivation to lead, honesty and integrity, self-confidence, ability to withstand setbacks, knowledge of the business and managing the perceptions of others in relation to these characteristics (Kilpatrick & Locke). If this were the case, then the property industry would be full of leaders!

Similarly, there were those who focused on charisma — that special quality of leaders whose purpose, powers and extraordinary determination differentiates them from others. This view sees leaders as people who are dominant, with a strong desire to influence, plenty of self-confidence and a strong sense of their own moral values (House). It is observed that many great leaders have boundless energy and enthusiasm and are both larks (those who are good first thing in the morning) and owls (those who operate best towards the end of the day and at night).

Yet more modern theories focus not so much on the "born to lead" models that do not hold out much hope for those who are not endowed with particular personality characteristics but on their behaviours in different situations which — theoretically at least — can be adopted and practised by all. These behaviours focus on task behaviours (which are directive — or, in extreme cases, autocratic) and relationship behaviours (supportive, coaching and nurturing). My "two person" leadership example above is therefore explained by these two different aspects.

Figure 3.2: Situational leadership — four leadership styles

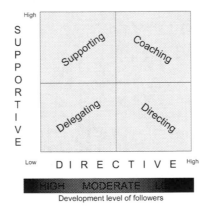

This suggests that you need to adopt different styles of leadership depending on the situation, for example:

- Dictatorial
  When a firm faces a crisis and there is no time to consult — use sparingly.

- Analytical
  When there is time pressure or threat and the right decision must be made quickly.

- Opinion seeking
  Builds team confidence and shows that you value people's views.

- Democratic
  Use regularly to empower team members and help strengthen their commitment.

I am sure that we all know good leaders and good managers, but it is rare to find both sets of attributes within the same person. Leadership accomplishes change whereas management focuses on maintaining equilibrium. Change management is a topic covered separately in chapter 9.

A leader creates the vision and a strategy to reach the vision (see chapter 2 on planning), whereas a manager uses a set of skills to work towards achieving that vision. Leadership in a partnership comprises those who lead and those who are led and their responsibilities to each other.

Avoid the trap of getting so caught up in conducting routine transactions with others that you neglect to inspire, persuade, influence or motivate them. And avoid the trap of thinking that what got you to a leadership position will keep you there. The Centre of Creative Leadership conducted research that found that success in early management positions is often associated with:

- Independence.
- Ability to control short-term results.
- Creativity.
- Ambition and high standards.
- Speciality strength.
- Being contentious — taking a stand.

Whereas in more senior leadership positions it is often associated with:

- Being a team player.
- Having longer-term strategic vision.
- Managing the creativity of others.
- Self-esteem.
- General management skills.
- Creating unity and cohesion.

In his book, *The absolute beginners guide to leadership*, Andrew J DuBrin provides his top 10 list for becoming an effective leader:

1. Understand that leadership is an influence process that includes inspiring, motivating and persuading others; creating useful visions and bringing about constructive change.

2. Develop self-confidence by achieving a small goal, and then continue to achieve progressively more difficult goals.

3. Develop your charisma. Learn to let your emotions show when you express yourself and remember people's names.

4. Become knowledgeable in some area of your business. Top leaders retain some important expertise, such as opening international markets.

5. Develop your communication skills — oral (spoken), written, non-verbal and listening — to the point that you are convincing.

6. Use a grab bag of influence tactics to positively influence people: persuasion, exchanging favours, ingratiating yourself and joking and kidding.

7. Lead by example. Be a model of how you want others to perform. Ethical leaders breed ethical followers.

8. Empower others by giving them the authority and responsibility for various tasks — exciting tasks, if possible. Empowerment helps people feel motivated and multiplies your effectiveness.

9. Learn to be a multicultural leader. The starting point is to convince yourself that although other cultures may be different from yours, they are equally valuable.

10. Be a great coach. Give people on-the-spot feedback, make suggestions for improvement, and offer encouragement and support.

This section concentrates on many of these ideas although chapter 8 on people contains material relating to motivating others, coaching and mentoring and other stuff to get your team producing effectively. Maybe we can break the concept of leadership down into its component parts, so here is my definition — but maybe it would be a good exercise for you to examine your idea of leadership — and then compare notes with your partners to build a combined view that meets everyone's expectations:

> "Leadership is the ability to influence others to act beyond routine compliance with procedures and daily tasks by demonstrating boundless energy and enthusiasm through good communication (including listening) and inspiring confidence and giving support through straight talking. It is the ability to motivate others to act or respond in a shared direction thus effectively a key dynamic force that motivates and co-ordinates an organisation to promote change directed at attaining agreed goals."

Now we can consider other different styles of leadership:

- The *visionary and transformational leader* is where you establish a vision as a starting point in building a glorious future for your firm. A transformational leader helps their firm and its people make positive and large-scale changes in the way they conduct their activities.

- A *servant leader* serves constituents by working on their behalf to help them achieve their goals, not the leader's own goals.

- The *flexible leader* is what expert Ralph Stogdill wrote about a long time ago: "The most effective leaders appear to exhibit a degree of versatility and flexibility that enables them to adapt their behaviour to the changing and contradictory demands made on them".

- An *empowering leader* enables his or her people to more freely exercise whatever power they possess, such as using their own expertise. It means releasing them from unnecessary real or imagined constraints and building their confidence to act on their own. *The Academy of Management Journal* indicates that empowerment has four psychological dimensions:

  1. Meaning.
  2. Competence.
  3. Self-determination.
  4. Impact.

- The *multicultural leader* has the skills and attitudes to relate effectively to people of different races, sexes, ages, social attitudes and lifestyles. There are many models to assist but Hofstede did some interesting work on the cultural differences between different nationalities. In essence, he shows that there are four dimensions on which there are significant cultural differences. These are individualist versus collective (noting that the US and UK are more individually oriented whereas those from many Asian countries are more group oriented), masculine and feminine (with the US being very masculine and countries such as Sweden being more at the feminine end), low power distance (for example, Americans feel comfortable with relatively junior people addressing senior people but in Eastern cultures there is a much greater respect for rank, seniority and hierarchy) and uncertainty avoidance (Western cultures being relatively tolerant of ambiguity whereas others need much more clarity and control). While most property partnerships reading this book are unlikely to be challenged too much in this area it is worth at least being aware that your leadership and management style may need to adapt as staff and clients become more diverse.

Jim Collins in *Good to great* has much to say about types of leaders. In fact, he offers a hierarchy to differentiate good from great leaders. He favours those who are quieter and perhaps a little more modest than most would often think. I think his idea of a level 5 executive has much resonance with the property industry:

Level 5   Level 5 executive
          Builds enduring greatness through a paradoxical blend of
          personal humility and professional will.

Level 4   Effective leader
             Catalyses commitment to and the vigorous pursuit of a clear
             and compelling vision, stimulating higher performance
             standards.

Level 3   Competent manager
             Organises people and resources towards the effective and
             efficient pursuit of pre-determined objectives.

Level 2   Contributing team member
             Contributes individual capabilities to the achievement of
             group objectives and works effectively with others in a
             group setting.

Level 1   Highly capable individual
             Makes   productive   contributions   through   talent,
             knowledge, skills and good work habits.

So which level do you think you currently fit? And how are you going
to move up to the next level?

But a leader in a partnership needs the support and a mandate
from his or her partner colleagues — and from all the staff to some
extent. So it is important that you align what you think your job is as
leader with the expectations of your colleagues. Some firms have a job
description for roles such as managing or senior partner. Others are
governed by what is written in the partnership deed. Some will
develop terms of reference that set out in broad terms the nature of the
role and what is expected within the next term — whether this is for a
year or for three years.

If you are being relieved of some or all of your client or fee-
earning responsibilities in order to tackle your management and
leadership role, then you will need to know how your contribution
will be measured. And this usually means what results are expected. It
is rare that you will be rewarded for simply keeping things ticking
over. So you must know what it is that the partnership expects of you.
At this point, I will point you back to chapter 2 because that is one of
the key ways that you can explore the problems and opportunities and
develop a plan and priorities for the firm and for your own action.

Many leaders find it helpful to have an "off-line" unofficial mentor
with whom they can discuss their ideas and challenges. Others find it
helpful to seek out others in similar roles in related organisations or

close by in the market. There are even groups where managing partners get together to discuss common challenges in a safe and confidential environment. It can be lonely at the top — so from where will you obtain advice, support and a sympathetic ear when you need it?

Checklist and actions

- What are your views about leadership?

- How do you need to change to transition from fee-earner or technician or manager to leader?

- What *is* your vision?

- What type of leader do you think you are? And what type of leader do you want to be?

- What are the particular skills or competencies or activities that you need to develop in order to be a great leader?

- What do your partners expect from you as their leader?

- How will your contribution be measured and rewarded?

- What is your mandate or terms of reference for your role?

- From where will you draw advice and support as a leader?

The helicopter view — strategic thinking

Too often, partners are thoroughly engrossed in the day-to-day stuff that surrounds them. When they demand change, it is for his or her small corner of the world and they give little thought to the impact of their requests or changes. That's fine for them, as their task is to manage their division or team (or just themselves). But a strategic thinker is someone who looks beyond the immediate situation and thinks about the causes and underlying problems and looks for a long-term solution rather than a quick fix. And a quick fix that may have catastrophic implications for others in the firm or even the firm overall in the longer term.

Strategic thinking is the ability to think in terms of how your actions help the organisation adapt to the outside world. A systems thinker understands how any change implemented in one part in the firm creates changes in the rest of the firm in the short and long term.

In some professional practices, it is hard to find the time to step back from the barrage of day-to-day pressures and detail and get some "heads up" time to look beyond the horizon and to see the big picture — to see what is important, rather than what is urgent, and to see what lies ahead and not just what is getting in the way.

Often, being a strategic thinker means asking those really tough questions that most people will hardly contemplate, yet alone have had time to try and formulate a coherent answer. If you are in any doubt about whether what you are tackling is strategic or tactical, the following checklist may help you:

Eight ways to distinguish between strategic and tactical decisions
(Adapted from Weitz and Wensley)

- Importance
    - Strategic decisions are significantly more important and more likely to have a profound or far reaching impact on your firm.

- Level at which conducted
    - Strategic decisions can only really be tackled by the owners or senior management of the firm. If someone more junior could manage it, then delegate it.

- Time horizon
    - Strategy is concerned with the medium and long term (at least a year ahead), whereas tactics focus on the short-term and immediate future.

- Regularity
    - Strategy is continuous, tactics are periodic.

- Nature of problem
    - Strategic problems are often unstructured and sometimes unique and involve considerable risk and uncertainty. Tactical problems are more structured and repetitive with risks easier to assess.

- Information needed
  - Strategies require large amounts of external information much of which is subjective and futuristic. Tactics depend much more on internally generated accounting or marketing research information.

- Detail
  - Strategy is broad. Tactics are narrow and specific.

- Ease of evaluation
  - Strategic decisions are more difficult to evaluate.

Here is a starting kit for acting like a leader: the 3Cs — confidence, credibility and communication.

Walk like a man, talk like a man — developing confidence, credibility and communication

I'm not being sexist or insulting to female leaders about the way that they walk. It means that you need to convey your confidence in the way you move — not just by what you say.

Interestingly, I usually teach those at the start of their career about conveying a confident persona so it is nice to be able to relay to those who are, presumably, at a more mature stage of life. It is not so much about actually being confident (although this helps) as getting people to perceive you as confident. You need to know what you are talking about — so the content is important, but the way you say things can be just as important.

Communication theory shows that only 7% of the meaning in an interaction is about the words you use. A massive 55% comes from the way you look: your appearance, posture and gestures (body language) and 38% from the way you sound: your voice. I try to help people remember this by talking about the dance, music and lyrics.

There are many ways to improve your confidence. Obviously experience and preparation are important. But there are other things as well. For example, you might ensure that your self-talk (how you speak to yourself in your head) is positive rather than negative. So, instead of thinking things like "I am no good at this really" or "I really don't think that this draft business plan is very good", think things such as "I have done my preparation and have some good ideas to contribute" or "This draft business plan presents some interesting information and offers some creative ideas — it will be a good foundation for further discussions". Other approaches encourage you to visualise positive situations or experiences and learn to remember these when you need a confidence boost — and this is based on the fact that your brain does not differentiate between the real situation or an imagined one, but your external non-verbal communication is driven by your internal state — whether this is created by the present situation or a remembered or imagined situation. This is the stuff of NLP (Neuro Linguistic Programming), which is a branch of psychology that focuses on the power of positive thought.

Other techniques might be to create a list of all your "assets" and "strengths" — areas where you have knowledge, can do things or are recognised as making a valuable contribution. It is important to remind yourself regularly that you have been put into a leadership role for a reason and, assuming that you trust the judgment of those partners who thrust you into the limelight, they must see your leadership potential even if you still need some convincing.

Your confidence is also boosted by self-efficacy (your confidence about your ability to execute a task) so it pays to be familiar with how to do simple things such as arranging meetings, sending emails, operating the main kit around your office and managing your work. For many partners, this means spending some time getting to grips with the numerous technological gadgets and systems that are there to help you to be more efficient.

You will be more confident when you become an action man (or woman) — someone with an action orientation is a person who wants to get things accomplished and decisions implemented. This will take energy and drive. Doing things and achieving results — even if you start small — will build confidence. So it is a virtuous circle.

Credibility

From psychology we know that the credibility of a communicator rests partly on his or her expertness and trustworthiness as perceived by the person on the receiving end of the communication (Hovland and Weiss). So, if you are someone who has grown up through the ranks, done great client work, generated plenty of business and managed a division or the firm for a while, then you will have some money in the credibility bank.

Credibility is reliant on the extent to which people trust you and this comes back to integrity. Sincerity is important here and if, as Groucho Marx commented: "If you can fake sincerity, you've got it made!". Closer to home, the management guru David Maister says: "Faking sincerity is a prostitute's tactic, not a professional's". But you need to be clear about your own personal ethical code. People need to know where they stand with you. If you have not given this much thought before, you should try and write down those things that are important to you (your values) and also some of your ethics (the behaviours you admire and those that you dislike).

Credibility also means that people must learn to trust you. And creating trust is about honest dealing and about doing what you say you will do — when you say that you will do it. Trust evolves when people feel that they know how you will react to things — in a consistent way. There is more information about the issue of trust in chapter 6 on selling.

Communication

We must not forget the importance of non-verbal communication — and emotional expression (for example, animated facial expressions, body posture, hand gesture) is a vital part of this. But you need control and having a leader who does major emotional outbursts is not what we want. But we don't want a cold automaton either. The most famous leaders were often passionate about their work — and their passion, enthusiasm and energy would be communicated in every aspect of their being.

So how you say things is important, but what you say is also critical — and your choice of words is vital. Linguistic style is a person's characteristic speaking pattern. It involves aspects of speech: the amount of directness, pacing and pausing, word choice and use of devices such as jokes, figures of speech, anecdotes, questions and apologies. Some people have a very formal style and others are more jovial.

Great leaders have the ability to inspire people by creating compelling visions of the future. To create great visions you need to use language evocatively. Analogies, metaphors and stories are powerful techniques. I still remember the faces of a room of surveyors when their leader was talking to them about the rather mundane topic of telephone skills. He said "Every time I pick up a ringing phone I think that this call is the one with the multi-million pound deal and so

I say "Hello" with a confident and welcoming smile on my face and enthusiasm in my voice". Everyone in that room now has a picture of their leader answering the phone in his characteristically enthusiastic and warm style.

Neuro Lingustic Programming (NLP) suggests that each person has a dominant style of language that is linked to their preferred sensory style. For example, a visual person will use the words "I see", "Look", "From my perspective" and an auditory person will use words such as "I hear what you say" or "Listen to my views" and a kinaesthetic person will say things like "I feel we are going along the right lines" or "My gut reaction". Great communicators will match the preferred style of the other person in a one-to-one interaction — thus promoting rapport, but addressing a group of individuals means ensuring that you use language that appeals to all types.

Good communicators have an ability to ask tough and insightful questions (see chapter 6 on selling for some thoughts on questioning and active listening).

Checklist and actions

- Have you allocated time to rise above the day-to-day hassles and think about the longer term and the environment outside your firm?

- Have you considered your strengths and weaknesses on the 3Cs — confidence, credibility and communication?

- How will you develop your self-efficacy — what practical things do you need to do to support yourself in your daily work?

- What are your personal values and ethics that will guide your behaviour in the small things as well as the big challenges?

- Do you have a distinctive linguistic style that conveys authority, energy, enthusiasm and humour?

- Are you careful in the choice of the words and verbal devices you use to create compelling images and drive action?

- How effective are you at adapting your personal style to the different people and situations you must deal with?

Where do I start?

This is probably the most important question of all. Hopefully, you will have read and used chapter 2 on planning so that you have an idea of what needs to be done, what needs to be approved by your partners and that you have a mandate to act.

Then you need to become organised. Some leaders of professional firms will be devoted full time to their management duties, while others will have to juggle with an active client portfolio. Whatever your situation, you need to decide how you are going to manage your time. It might be useful to keep track of how you use your time for a couple of weeks to see what can be delegated or dropped to free up some time. You can download a management diary from my website that enables senior partners to think about their time use over the period of a week and this usually provides useful insights into how you can work more effectively.

Time management is a whole new topic. You might do well to look at Stephen Covey's *Seven habits of highly effective people* (the identification of urgent versus important activities is crucial) and you might look up some of the most common time management techniques, such as goal setting or rocks, stones and pebbles. These are covered further in chapter 9 on change management.

Wanna be in my gang?
Picking a management team

You will need a small team around you to help you lead. Too often the choice is restricted to those who share the most equity — and probably the same style, values and views. But you need diversity in your team. People with different functional expertise, those with different knowledge and outlooks, and with different styles. There is a well known assessment of team behaviour by Belbin, who suggests that a good team needs a balance of the following:

Shaper	Task focus
Chairperson	Leads
Plant	Ideas
Resource Instigator	Gets things
Monitor Evaluator	Checks/asks questions
Team Worker	Promotes harmony

Company worker   Follows rules
Completer Finisher   Does the job

Merril looked at this from a slightly difference perspective:

Table 3.3: Personality types of managers

Analytical	**Commander**
(Thinking)	(Action)
Prepare well in advance	Be brief/to the point
Be clear and rational	Avoid chat/social pleasantries
Avoid emotions	Stick to the business at hand
Create action plans	Mention objectives and results
Amiable	**Expressives**
(Relationships)	(Intuition)
Take time	Entertain and stimulate
Learn background	Be lively and enthusiastic
Ask for opinions	Ask their opinions
Reassure them about others	Remember the big picture
	Don't get lost in the detail

Source: Merril

In addition to the core management group, you may wish to create other groups to tackle specific projects or issues. As well as selecting the appropriate members, you will need to clearly identify what you expect the group to achieve, the scope of their operations and the timescales and resources that they have available. It should be made clear how you expect the group to operate, and how they record their time and how they report progress — on what time basis and to whom.

Meetings, bloody meetings

Sometimes, a leader will find him or herself asked to attend many meetings. To participate in the work that has been delegated to particular committees or groups. But be careful with your time and insist that if you are to attend you need to know the reason, what materials you should review in advance and the specific areas on which you are being asked to contribute. Remember that you do not need to attend the entire meeting — you can agree a particular time slot to attend. Another trick is to ask your secretary to only book 15 minute slots in your schedule for attendance at meetings. When people know that you will only be around for a short time, they will work harder to prepare and to be clear what they want to achieve with you in the short time that you are available. Another tip — particularly valuable to those who have to cover a network of offices — is to agree to teleconference or webcam into part of a meeting at a particular office.

Some firms rotate the location of their key management meetings so that the senior people are guaranteed to visit every office at least once a year. In addition to spreading the travelling load, your presence in the branch office will be appreciated and it gives you an opportunity to wander round, chat to the people there and to get a feel for how things are going.

Of course, as the leader you will be responsible for getting the most out of your team(s). Remember that it will take time for a new group to go through the stages of forming (inclusion) — storming (assertion) — norming (co-operation) — performing (producing results). And perhaps here is a good time to talk about the way to act when chairing a meeting (Rackman and Morgan):

- Initiating
    - Propose ideas and suggestions to encourage people to talk.
    - Build on their ideas and get others to contribute their views.

- Reacting
    - Support other members and their ideas and prompt debate.
    - Manage disagreements between people — do not let conflict interfere with purposeful discussion.
    - Prevent people becoming too defensive or aggressive, but instead encourage people to focus on the merits of the idea(s) in a more objective way.
    - Avoid people blocking ideas or debate — encourage people to state any potential difficulties or constraints instead.

- Clarifying
    - Open the debate by asking questions.
    - Test everyone's understanding of key points.
    - Summarise the key points raised and agreed decisions and actions.
    - Seek more information when required.
    - Give information when it is required.

- Controlling participation
  - Shut out those who are dominating discussion occasionally.
  - Include everyone at some point in the meeting.

Sometimes it is a good idea to appoint someone else to chair the meeting. That leaves you free to listen more carefully to what is being said, think about the motivations of the various people as they talk and step off of the fence that often chair people find themselves on so that you can give an honest view. It is also a great development opportunity for those who find themselves suddenly thrust into the limelight!

Naturally, you will ensure that minutes are taken at meetings. For short, informal meetings this might just be an email confirmation of the one or two action points that were agreed — along with timescales. For an important partnership meeting, you may want to record some of the detail of the discussion and the options explored (and reasons for adoption or rejection) if you want an accurate record of what was considered or if the minutes will communicate the essence of the discussion and different viewpoints explored to those who were unable to be present. As well as recording what was said and agreed, providing an aide-memoire so people know what they must do and a mechanism by which progress can be measured, minutes also provide an important communication device — enabling those outside the decision-making circle to keep in touch with what is happening and why. Many firms will have all minutes kept in a central public file or on the intranet so that the authorised group can easily review decisions and progress.

Checklist and actions

- Get organised — decide what time you have available and how you want to use it.

- Decide on your priorities — as dictated by the business plan or by other discussions. Identify the way in which you will tackle the various projects and tasks but leave time to deal with smaller, urgent matters as they arise.

- Give yourself time on a regular basis to review the plan and the big projects so that you can ensure that everything is moving in the right direction and that you do the necessary "How's it going?" chase calls.

- Consider the various teams you need and the composition of those teams — and don't limit your choice of members to just the partners.

- Give groups clear terms of reference as to what they should be doing and the results that they need to produce.

- Develop your chairing skills and engage others in the activity.

- Brief your PA on how to allocate time for meetings in your diary.

- Ensure that your secretary reminds you to prepare for imminent meetings and allocates the time to do this in your schedule.

- Evolve different styles of taking and using minutes.

- Ensure that you have a programme of regular visits around all your branches, offices and teams.

Power, persuasion and influence

Chapter 9 is devoted to change management — how to look at the big picture of where you are now and where you want to be and move the entire firm in a new direction. But on a day-to-day basis, as a leader, you will need a considerable toolbox of abilities to achieve your goals through using your power, persuasion and influence skills. Below, I have selected a few of my favourite techniques as, again, it is an immense subject in its own right.

- *Role model* — "Do as I say, not as I do" simply does not wash in a professional environment. For many years I have observed young professionals mimic or adopt those behaviours of their seniors and leaders rather than do what they are told to do. So if you let your partners do as they like, then you can be sure that their subordinates will follow suit. This is vitally important when it comes to change management programmes, because you can implement change so much faster if all the partners adopt the new behaviours first. So training the juniors while the partners carry on as before is a waste of money.

- *Benefits* — People are more likely to do as you want or ask them if they can see the personal benefits of doing so. So, in addition to linking your and the firm's aims to their personal aims you must be able to see things from their perspective and understand their particular motivation (see chapter 8 on people).

- *Champions* — Champions are those who provide political support to your ideas because they feel a personal allegiance to you and thus become the champion of your ideas. This is linked to the idea of power people below.

- *Expertise* — Being a technical expert is usually an effective influencing tactic but if there is too much content in your role of technical expert it could block perception of leadership of more general and strategic matters.

- *Inspiration* — An inspirational appeal is influencing another person by drawing on a strong emotion. This is often done in conjunction with the work on developing a powerful and compelling vision of the future (see chapter 2 on planning).

- *Obtain their input* — Communication is discussed in many parts of this book — particularly in planning and also in people. But never forget to gain the views of others as they will feel valued and involved in any subsequent decisions. Furthermore, extending the ideas in the people section, you might consider implementing a 360 degree review where you obtain the (anonymous) views of those you manage on your style.

- *Personal power* — There are three main ways in which we have and use personal power. The first is presence where the main tool is physical attraction although this can be intimidating. Then there is authority, which is use of expertise or knowledge and a common approach for professionals, although the downside is that often people who use this always want to be right. The final element is impact — the ability to create change through asking incisive questions.

- *Power people* — Having the ear of the power partners will be a help. But you can use those outside the firm too — by developing your network of useful personal contacts and gaining their support for internal initiatives. I have seen this done very successfully when leaders involve client speeches or contributions from industry events (and even journalists) at internal events.

- *Success stories* — Whereas it is common to focus on the negatives, or on those things that went wrong or are yet to be achieved, it is immensely powerful to talk about success and to provide examples of successful "war stories".

- *Visibility* — Many leaders have to take on the chairman role of promoting the firm in the external marketplace with public relations and talking to the press. Being seen and heard and taking credit for your successes is an important way to increase your influencing skills. But care must be taken that you are not seen as blowing your own trumpet and ensuring that your team is recognised also for its contribution.

If your role extends to negotiating, then a good starting point is "Getting to yes — negotiating an agreement without giving in" (Roger Fisher and William Ury) where the focus is on "principled negotiation produces wise agreements amicably and efficiently". It means that as a leader or manager you may have to adopt a slightly different approach to negotiating than you do for property transactions. They stress the need to achieve win:win results if you:

- Remain open to persuasion by objective facts and principles.
- Avoid stating your position at the outset.
- Do not bargain over positions — as the more attention that is paid to positions, the less attention is devoted to meeting the underlying concerns of the parties.
- Negotiate the substance but also on the procedure for dealing with the substance (usually without conscious decision).
- People — separate people from the problem.
- Interests — focus on interests, not positions.
- Options — generate a variety of possibilities before deciding what to do.
- Criteria — insist that the result be based on some objective standard.
- Analyse the situation. Plan the outcome. Promote open discussion.

There is more material on influencing in chapter 6 on selling. I think Dale Carnegie's book *How to win friends and influence people* is one of the best primers. He offers some fundamentals on handling people such as to avoid the 3Cs (criticise, condemn or complain), give honest and sincere appreciation and "arouse an eager want". He offers six ways to make people like you:

- Become genuinely interested in other people.
- Smile.
- Remember a person's name is the most important thing.
- Be a good listener — encourage others to talk about themselves.
- Talk in terms of the other person's interests.
- Make the other person feel important and do it sincerely.

He offers a toolbox of techniques to encourage people to your way of thinking. Some of his best pieces of advice are: the only way to get something from an argument is to avoid it, show respect for others' opinions (never say "you're wrong"), admit it quickly and emphatically if you are wrong, get the other person saying "Yes, yes" immediately, let the other person do a great deal of the talking, let the other person feel that the idea is his/hers, try honestly to see things from the other person's point of view (that important empathy concept again), be sympathetic with the other person's ideas and desires, appeal to nobler motives, dramatise your ideas and throw down a challenge.

Checklist and actions

- Build your internal network of power people and champions.

- Extend your external network of influential people — especially beyond the property sector.

- Ensure that your partners are good role models for the behaviours you wish to encourage.

- Develop your understanding of what is important to, and drives, those around you.

- Illicit people's views and ideas on the way forward in general as well as specific projects.

- Build a stockpile of success stories to encourage and inspire others.

- Recognise and promote the success stories.

- Develop your toolbox of persuasion and negotiation techniques.

Delegate don't abdicate

This must be one of the most challenging tasks for a leader — especially if he or she is exceptionally good at a wide range of tasks. There is the danger that they will "hang on" to too many actions which cause delay in overall progress. Micromanagement is the close monitoring of most aspects of group member activity by the leader or manager. But you can't be an absent boss either — that's abdication. Delegation entails assigning duties, granting authority and creating an obligation. For it to be effective, all three components must be present.

The book *The Art of Delegation* by Ros Jay and Richard Templar sets out the following simple checklist each time that you delegate:

1. Review the task and set the objective
2. Decide to whom to delegate
3. Set parameters
   a. Objective
   b. Deadline
   c. Quality standards
   d. Budget
   e. Limits of authority
   f. Details of any resources available
4. Check they understand
5. Give them back up

And remember that, as Minzberg wrote, professionals and knowledge workers respond better to inspiration than supervision.

Amaze as an ambassador

Another one of the duties that they slipped into the brief when you took on your leadership role was that of being a spokesperson, PR guru and generally amazing ambassador for the firm. Huh! This means attending an incredible array of events both within the firm and externally where you must slap on a smile, forget the mountain of work back at base and go into "what a great firm we are" mode.

It is important that in whatever you say you keep in mind the key messages that you wish to convey about the firm — so we are back to having a good plan and a clear strategy. The strategic marketing section of the book (see chapter 4) will also be useful to you here.

As you grow in experience as a leader, you will become adept at making impromptu introductions to the firm or being the master (or mistress) of ceremonies at key partner, staff, client and community events. While nerve-wracking to begin with, it soon becomes second nature but you must guard against churning out the same old dull platitudes that everyone else does.

Sometimes, you will be called upon to act as spokesperson for the firm on important issues with the media. This requires a new set of skills. And one where it is best to obtain professional help if it is a new or worrying situation for you. I might be so bold as to refer you here to one of my previous works *Media Relations in Property*, which I co-wrote in 2006 with a leading residential property journalist — there is plenty of advice in there on how to deal with tricky reporters and front-of-camera experiences. There's some more information on media relations in chapter 5 on tactical marketing.

Pick up on the particular interests or language of your audience, link to topical activities, use humour and convey warmth. You are the living embodiment of your firm's brand or reputation and it will survive or dive according to your moods and performance. So be the firm's best ambassador.

Bring in the experts

Sometimes you might need to bring in external experts because you need an outside, dispassionate view or access to expertise that you do not have or an external interventionist or facilitator.

While consultants can be expensive, they can — if chosen properly — save a lot of time, money and grief. How many projects do you know that have gone wrong because they started off in a half-cocked manner with a bunch of amateurs thinking that they know best but then having no time (or expertise) to enable them to rise to the challenge?

Before you commission an expert you need to do a number of things. Write down a brief of what you want — and what you do not want. Think about the deliverables or results that you need. Sometimes you want a consultant to provide you with a plan and other times you want the consultant to do some or all of the implementation. Then be realistic about the time and money you are prepared to invest. Never forget that most consultants will be a heavy drain on your time and the management agenda if they are to do their job properly. Then obtain input from others about the brief — gaining commitment to the external consultant idea is important in a partnership.

Then you need to find the right consultant. This is easier said than done. They need to have the relevant expertise and hopefully some experience of the property sector or one where there are clear parallels. And they need to know how to work with a partnership.

Then you need to do some research on the internet and ask around for personal recommendations. Spend some time with the different consultants understanding what they have to offer, how they work, what they cost and, most importantly, whether their style "fits" with your partnership.

Checklist and actions

- Identify those things that you do not have to do yourself and those that can take some of the load off your shoulders.

- Develop your delegation skills — and ask others to indicate how you can improve.

- Think about the key messages about the firm that you want to convey on all occasions.

- Practise making impromptu speeches for internal and external, informal and formal events.

- Book yourself onto a course to develop your skills and confidence at dealing with the media.

- Review your effectiveness as a great ambassador for all parts of the firm.

- Recognise when you don't have the expertise or resources to tackle an issue.

- Prepare a brief before attempting to recruit outside experts to help you.

- Develop your network of contacts so that you have access to people who might know appropriate experts and outside helpers.

Banging heads together — managing conflict

It is inevitable that as you attempt to lead your firm into unchartered waters, there will be tension and perhaps conflict. That conflict inevitably involves people but the cause may be about personalities, projects or priorities — sometimes even a clash of aims or value. Some research suggests that dealing with conflict can take up to 20% of a manager's time.

If you are sitting there smugly thinking that there is little conflict in your firm, then think again — maybe there is but you are not the sort of environment where differences can be aired in a constructive manner. While harmonious partnerships are pleasant, there are dangers in "group think" (ie, "a deterioration of mental efficiency, reality testing and moral judgment in the interest of group cohesion. The group tries so hard to get on with each other that they may accept outrageous ideas", Janis) — where you all think alike or where no one provides valuable checks and challenges.

Inevitably, if there is conflict among others then you must confront it — sensitively and with tact rather than combative and abusive. Here are some steps:

1. Recognise that the conflict warrants action. Will it work itself out or continue to cause problems? Some things are best left alone to let time heal.

2. Scope the situation. Do you have enough information? And does that information explain both sides of the situation?

3. Make the confrontation. But choose your moment carefully. Avoid a confrontation when the person is busy with something else or heated up from a difficult exchange.

4. Determine the cause of the conflict.

5. Develop approaches to reduce the conflict and apply them. In essence there are some obvious strategies — withdrawal, smooth, compromise or force — and more creative approaches.

6. Monitor progress.

Often you can reduce tension in a situation by disarming the person — by starting out showing empathy and agreeing with them. Then an indication that you are not so concerned with apportioning blame as finding a solution and avoiding the situation in the future. That usually calms things down a bit. Ideally, you want to achieve a win-win situation — where each party feels that they have come out of the situation a winner. Sometimes you can reframe negative things by considering the more positive aspects of a situation — the Monty Python "Always look on the bright side of life" approach.

Mr (or Ms) Fixit — creative thinking and problem solving

Many great leaders are graced with the ability to look at the big picture — or all the component elements — with the same information that everyone has, but to come up with an original and creative solution Here we see the leader showing that ability to "think outside the box". How do they do it?

Some people say that a true expert is someone who has at least 10 years' experience in their field and has therefore effectively assimilated and organised all the relevant knowledge so that their thinking processes are more at a subconscious level and that their brain is more able to make "intuitive leaps" than those with less experience. There is also an important role in confidence here too — many people will have great ideas, but only confident people will dare to share and risk having their thoughts possibly shot down in flames with the attendant discomfort and reputation risk.

The nature of education and training for most professionals means that they will usually rely mostly on the left and logical side of their brain and their rational, analytic thinking abilities. Yet creative people manage to access and also use the right-hand side of their brain — which is more concerned with emotion and language and creativity. Usually it is not obvious that they are doing this — but they rely on their gut feeling or their intuition to help them move to a creative answer.

Also it should be remembered that your people are a vital source of information and good ideas. But they are unlikely to share those ideas if they feel the environment is unsafe. Nancy Kline, in her book *Time to think* outlines the 10 components of a thinking environment and illustrates how the dominant, masculine style is so different:

Table 3.4: Ten components of a thinking environment

Thinking environment	Male conditioning
• Listen	• Take over and talk
• Ask incisive questions	• Know everything
• Establish equality	• Assume superiority
• Appreciate	• Criticise
• Be at ease	• Control
• Encourage	• Compete
• Feel	• Toughen
• Supply accurate information	• Lie
• Humanise the place	• Conquer the place
• Create diversity	• Deride differences

Source: Nancy Kline

This would suggest that we review the dominant culture in the firm — which may necessarily be quite male conditioned in view of the nature of the property market and/or work that we do — and consider how we modify that culture for the purpose of internal meetings where we want staff to feel secure enough to contribute their thoughts and ideas. Freeing up the minds of your staff provides a far greater pool of potential ideas that relying solely on your own.

A good starting point is to have some frameworks to help you work through problems in a structured way. All professional people like processes, how's this for a problem-solving approach (a more detailed version is available from Grundy):

1. Identify the problem.
2. Clarify your problem.
3. Analyse the cause.
4. Search for alternative solutions.
5. Choose one or more good alternatives.
6. Plan for implementation.
7. Establish contracts and commitments.

On a day-to-day basis, I often use the much shorter 4D approach to help other people solve their own problems by coaching them through the following stages:

- **D**efine the problem in a single sentence.
- **D**escribe:
  - the situation (eg, time, resources);
  - the people involved; and
  - you.
- **D**evelop options (and identify one priority issue from each option).
- **D**ecide on the appropriate action.

But there is more to creativity than process isn't there? Jane Henry, an expert in the area of creativity, suggests a four-stage model of creativity:

- Positivity
  - A habit of seeing problems as opportunities.
  - Rapid recovery from setbacks.
  - Tolerance of criticism.
  - Unwillingness to let blockages impede progress.

- Playfulness
  - Drawing on childhood resources.
  - Taking (calculated) risks.
  - Using fun and humour in thinking.
  - Being flexible in thought and deed.
  - Feeling comfortable outside mainstream thinking and action.

- Passion (purpose)
  - Driven by a consuming purpose.
  - Obsessive will to achieve goals.

- Persistence
  - Try something different until you succeed.

Other researchers have identified other aspects of creativity — such as a tolerance for ambiguity, not being constrained by conformity, imagination and visual thinking, questioning skills, identification of gaps in knowledge and being open to new experiences. Perhaps, in some meetings, you might suggest that a more creative approach is required, and think about these four stages and different attributes and "give permission" to people to be more positive and playful when considering problems and issues.

I often talk to people about the value in alternating between divergent and convergent thinking. Typically, we converge — we try to find patterns and trends and to summarise. In divergent thinking, we allow our minds to wander more broadly and go off on sometimes implausible tangents to open up the field of thinking. Occasionally converging means that you can focus on the one or two nuggets that were revealed during the divergent phase. But it takes skill and confidence to lead a group of people through the unchartered and sometimes scary waters of real divergent thinking.

A fun tool you can use at workshops and meetings is to allocate specific roles to different people to force them to think about a situation in a particular way. I like Edward De Bono's "thinking hats" approach:

Table 3.5: Edward De Bono's "thinking hats"

Colour/role	Focus
White	Facts, figures, information
Red	Intuition, feelings, emotions
Black	Judgment and caution
Yellow	Logical positive — why, benefits
Green	Creative, alternatives, ideas etc
Blue	Overview or process control

Source: Edward De Bono

The good old "write down ideas on yellow stickies" approach is good at generating ideas too. And it can be useful to arrange or group the results onto a white board that will act as a memory aid and perhaps provoke further ideas after the session. Mindmapping or brainwriting (where you write an idea on a piece of paper and pass it around for someone to adapt or add to the initial idea) are well-known methods to encourage groups to be more creative. With senior partners, I find the superheroes approach is rewarding — asking people how they might tackle a situation if they were "Superman" or "Storm" generates some interesting insights and can unleash the inner champion. I also admire the use of drawing as an aid to the creative process, which is described brilliantly in *The back of the napkin — Solving problems and selling ideas with pictures* by Dan Roam.

Generating good ideas is one thing, but working out which of them is worth pursuing is another. But we are probably all much better at taking ideas forward than we are at generating new ideas and your assessment and project management skills (see chapter 9 on change management) will come to the fore here.

Checklist and actions

- Consider how much time you devote to resolving conflicts — is it reasonable or are there systemic problems (such as the lack of an agreed vision, plan or priorities) that you need to address?

- Be on the alert for "group think" — if your partners are always agreeing with you then there is a danger that you are not open to new ideas or ways of thinking.

- Take positive action to create a more thinking environment.

- Seek ideas and input from all members of the firm.

- Identify problem conflicts and take early action to address them — don't leave things to fester or to build up to explosions.

- Develop some problem-solving approaches that you can apply in day-to-day situations.

- Find and practice using some new tools to promote more creative thinking.

Don't turn a crisis into a catastrophe

Sometimes things go wrong. A crisis is a major, unpredictable event that carries with it potential results of enormous negative consequences. Sometimes it is the fault of an individual in the firm and sometimes it is a client or a market change. In the heat of the moment it is easy to panic. But a crisis needs calm and clear thinking. While all around you are panicking and wringing their hands and shouting at each other it is important that the leader takes control, considers the situation and the possible remedies, reassures everyone and then puts an effective action plan into place. The leader must lead his or her team through the crisis.

Jim Collins, in his book *From good to great* urges leaders to create a climate where the truth is heard. He advises:

1. Lead with questions, not answers.
2. Engage in dialogue and debate, not coercion.
3. Conduct autopsies, without blame.
4. Build in "red flag" mechanisms.

He refers to "the stockdale paradox" where you must retain faith that you will prevail in the end, regardless of the difficulties while at the same time confronting the most brutal facts of your current reality, whatever they might be. A sort of optimism combined with realism.

Yet an axiom of leadership theory is that when a group faces a crisis, a directive and forceful style of leadership works best. John Ramee offers the following guidance for when you must move into "crisis management mode":

1. Stay cool under pressure.
2. Avoid the quick fix that will hurt the organisation in the long run.
3. Seek new information.
4. Revise strategies.
5. Have a centre of authority.
6. Act quickly and decisively.
7. Trust your intuition.

In crises, communication becomes even more important than usual — if this is possible. Establishing control quickly, by perhaps directing all calls and enquiries and decisions to a central point is a good start. This way, the leader can see all the elements arising and attempt to ensure that all questions are answered consistently.

Sometimes on the television you will see the head of the company at the site of the "disaster" looking concerned but calm and indicating that he (or she) is doing all that they can to assess the situation and resolve it. This is good PR. But it is also good leadership. Compare this approach to those organisations whose leader disappears from view at the first hint of a problem, leaving his poor staff to make empty statements and leaving everyone in the dark.

If your firm is really in dire straits, take a look at the section on turnaround strategies in chapter 2 on planning.

Checklist and actions

- Invest some time in thinking about how you would deal with a crisis if it arose — what would you do? How would you act? What systems do you need in place?

- When a crisis arises, take time to gather and assess the facts and the risk.

- Think before acting prematurely.

- Adopt a more directive approach to leadership.

- Ensure that your communication in a crisis is really controlled but includes all the various stakeholders (partners, staff, clients, industry, community etc).

- Have central systems and procedures in place to deal with a crisis should it occur.

- Recognise if the crisis is really critical and be prepared to seek external expertise and advice if a turnaround approach is required.

Case study: Henry Adams

Richard Williscroft started the residential estate agency Henry Adams in Chichester in 1992 with a negotiator and a secretary in a small office. Today, Henry Adams is a highly respected multi-disciplinary practice offering residential and commercial sales and lettings, fine art auctions, valuations, waterside and agricultural properties, surveying and planning, with offices throughout Surrey, Sussex and Hampshire. He talked to me towards the end of 2008 as the property market slumped to even lower depths.

"For the first 12 months I wondered more than once what on earth I was doing and just when and if Henry Adams was going to succeed! Few people came in the office, so the only way to win business was to go out and get it. I aimed to cover a complete cross-section of properties and whenever an opportunity presented itself, we were there to seize it. I was, and still am, very fortunate in having utterly reliable and dedicated staff whom I know will always hold the fort and go that extra mile for our customers, be they vendors or applicants, and in the second year we went into profit. After four years, we became the market leaders in Chichester, and two more members of staff joined our team.

Developing people who develop the business

I was approached by a national agricultural firm with an office in Petersfield, and that allowed us to open a second residential office there. The location was fairly tucked away, but the live wire who managed the residential side was, and still is, a human dynamo. There is no magic in this business, you just need lots of energy and a really positive attitude, and the office was a classic success story. After four years, I made him a manager in Chichester and he subsequently set up, from scratch, our lettings business. Since then, he has moved on to our Bognor group of offices and set up the lettings business there. I'm delighted to say today he is one of our equity partners.

As with any relationship, the one between employer and employee is a two-way thing; looking after our staff and rewarding their loyalty and dedication with opportunities ensures they stay with us. Several of our staff have been with us since the very early days and bring a wealth of experience to our team. In fact, the vast majority of the staff in Chichester have been with me for well over 10 years, an unusually long length of service for our industry. That sort of loyalty cannot be underestimated.

Growth lessons learned

We've always listened to customer feedback and market conditions, and tried to stay one step ahead of the market and current trends when we expanded. I believe it has been a great advantage to us that Henry Adams covers such a variety of properties. Not only does that not pigeonhole us as a certain type of agent, but also means we have the flexibility to move with the times and focus our efforts according to the market and economy.

There have been several mergers, although they may have been seen more as takeovers, and they have always worked well. People joined us with specific roles and with a sense of identity in our brand, and regular conferences and the team spirit which runs through Henry Adams helped cement our relationships.

With our growth came the need for head office functions such as human resources, IT support and marketing, but the downturn in the property market meant we needed to look at our firm's structure in a different way. We demerged various parts of the business and created separate LLP's. Everyone went back to the coal face, rolled up their sleeves and got stuck in. It's back to basics and fee earning! Interestingly, it's when life is tough that we realise just how loyal our staff are, and the time invested in creating a good working environment in the office pays off. The old Dunkirk spirit has come to the fore, morale is high and I genuinely believe our staff are proud to be associated with our brand.

Several of our partners have moved on to a franchise model, so that each LLP can control its own destiny, and as a result of this franchise model, Henry Adams still has more boards up than anyone else in this part of the country.

Advice for others?

Marketing is key. When you start out, you have to be led entirely by marketing initiatives rather than accounting matters. Take a deep breath and put finance second place — be proactive and get out there. Your marketing cannot be too clever, and where we have scored is having a very strong reputation for honesty and fairness. Estate agents all start with negative PR and bad press, so you have to work extra hard to contradict that and show you are different from other agents.

I believe our strongest strapline is: "Not all agents are the same", and we aren't the same. In the same way that all Henry Adams staff are treated equally and with mutual trust and respect, we form good relationships with our clients and feel equal to them rather than their agents. There is always a very positive attitude in the office, and this working environment makes our staff comfortable and at ease with the clients. Our very high repeat business and recommendations statistics bear that out, with 97% of our clients saying they would use us again. But reputations are lost far quicker than they are won, and it only takes one mistake to taint our high-quality image. Because of this, we have invested heavily in our staff and their training and were delighted to get the Investors in People award this year.

At the end of the day it's all about people. The connection that is made when the phone is answered is vital, both in terms of customer service and reputation. A lot of very hard work has gone into Henry Adams over the years that the public would never see, but that counts for nothing if the public don't come in. Our staff are our strongest asset."

Case study: Chase & Partners

Background and philosophy

Chase & Partners is a niche firm of chartered surveyors and chartered town planners specialising in the retail and leisure property markets. It has one office in St James's, London, but covers the entirety of the UK in terms of its activities.

There are seven partners of the LLP and a total staff compliment, including partners, of 20 individuals with a wide variety of skills across several areas of business. These include occupational agency, investment, development consultancy, professional services and valuation, town planning and local authority consultancy.

Turnover varies depending on the state of the market and part of Chase & Partners' business is transactional which produces higher returns when markets are strong, but in weaker markets its professional consultancy and planning activities come to the fore albeit with lower turnover profiles.

The corporate strapline is: "a personalised, solutions-based approach". Against this background, we attempt to combine experience with innovation. It is a fact that no two projects are the same, each client has different requirements and at the end of the day we are judged on results. Consequently, strong relationships with existing clients is a key feature of our business approach and this is coupled with our exposure to the market at the cutting edge, where we are constantly coming across new business and new potential clients.

Business planning

Despite being a small niche practice, business efficiency and success is run on the same lines as some of the biggest corporates in our sector. Every year, the partners organise a business planning meeting, away from home, and feed into a three-year rolling plan. This takes account of past business and performance, client activities and requirements and current and estimated future market conditions. The resultant business plan not only enables us to maintain a tight efficient accounting system, but also provides a focus and objectivity that allows us to move forward with all partners and staff heading in the same direction.

The business plan is complimented by monthly management accounts, which analyses our income and costs in terms of actual, forecasts and budget criteria. These accounts therefore are our eyes in assessing how we are doing in the market and to make sure we adapt to changing economic conditions, so as to ensure the health of the practice.

Once a month, business meetings are attended by all surveying staff, where each individual is encouraged to participate reflecting on existing instructions and identifying new ideas. Each department also meets once a week to look at their own activities and to ensure each team is co-ordinated in terms of their activities and approach.

To date, Chase & Partners has for the past 10 years been in the top quartile of the *Estates Gazette* survey in terms of earnings per head and profitability. Although much of this is down to the expertise of the individual members of staff, we believe the success also reflects a clear business plan with set objectives and an ongoing monitoring process of our performance.

Relationship with our clients

Being a small niche practice, it is not possible to deal with everybody. We therefore focus carefully on our client base although make sure we are in regular contact with all players within our market. The gathering of intelligence, which is up to date and correct, tends to come from regular communication with people who are active in the market and making decisions rather than the analysis of reports, data and statistics, which inevitably tend to lag behind the market and offer a view in the rear mirror rather than an outlook through the front windscreen.

In support of our knowledge and expertise through the marketplace, we prepare regular market briefings and case studies, which are available on our website. Since Chase & Partners was founded, we have also prepared an "End of Year Retail Report" in December of each year, giving our assessment of the market over the previous 12 months, our considered opinion of current market conditions and our projections for what may be in store for the retail and leisure property markets over the next 12 months. Again, these documents are posted on our website. They help to provide a discipline for those who write the reports and make us all sit back and consider what we do more carefully.

The report as an anecdotal up-to-date market base commentary tends to find particular favour with our client base, given that we do not pretend to compete with the more statistical research-based reports prepared by some of our larger competitors.

Staff training

Having staff who are up to date and motivated relies on ensuring that they feel able to compete within the market and have confidence in their knowledge. We are therefore supporters of personal development planning, continuing professional development and life-long learning. As well as complying with RICS and RTPI ongoing training requirements, our staff regularly attend external conferences and Chase & Partners hosts internal training sessions on a regular basis, primarily covering new issues on professional and valuation matters as well as changes in planning policy.

Given that the partners in Chase & Partners are some of the best known individuals within their respective marketplaces the passing down of their knowledge to new members of the practice is essential. It also allows the partners to learn from each other and every day we tend to learn something new. It is a key requirement that members of Chase & Partners specialising in one area are nevertheless aware of current initiatives and market issues in other sectors, as this aids joined-up thinking and knowledge in terms of our overall position in the market. Stop learning and you stop being effective and will not add value to your clients' activities.

Record keeping

The keeping of internal records is a vital activity even for small firms. Up-to-date databases, including contact lists, details of transactional evidence, business planning and performance valuations and simple things such as the keeping of up-to-date diaries, helps improve efficiency and ensures that the wheel does not have to constantly be reinvented. This requires hard work and commitment from each individual within the practice but the rewards from maintaining quality databases far outweighs the tedium of keeping them updated and relevant.

Summary

Chase & Partners takes the view that "leaders think and talk about the solutions, whereas followers think and talk about the problems". We produce honest and straightforward advice based on knowledge and understanding in the market. We will tell it as we see it but always in a constructive and supportive manner. We regard ourselves as partners in a team effort where the emphasis is on communication and a clear eye on the objectives as set.

It is not possible to advise on property unless you have seen it and we make sure that our advice is based on not only our understanding of the property but also all the circumstances that surround it. We also know that our job is to see our clients rather than expect our clients to travel to us. Simple courtesies are usually the best way to establish your credentials and move forward in a difficult and challenging business environment and where the rewards have to be won and not expected.

Strategic marketing (marketing strategies)

Roadmap

- Strategic versus operational marketing?
- Why marketing your practice is different to marketing a building
- The marketing audit — strategic analysis and research
- Business sector analysis
- The crown jewels — key client analysis
- Where does it come from? Source analysis
- Stars, dogs and cows — portfolio analysis
- Setting SMART objectives
- Strategic choices — competitive advantage, target markets and priority services
- Finding your niche — segmentation strategies
- Innovation and new service development
- Pricing
- Spot the difference — commodities and brands
- Decisions, decisions — what makes a good marketing strategy?
- To plan or not to plan? Simple marketing plans
- How much? Budget questions
- Make or buy? Acquiring marketing expertise
- Keeping your eye on the ball — monitoring

"Selling focuses on the needs of the seller.
Marketing focuses on the needs of the buyer."
Theodore Levitt

"Marketing is the management process for anticipating
and meeting client needs profitably."
Chartered Institute of Marketing

Strategic versus operational marketing?

This book addresses marketing in two separate chapters. Most of the time, the property industry focuses on tactical and operational marketing — the actual things that you need to do to promote your practice, your service or your client's buildings and developments, yet the effectiveness of your operational marketing (or marketing communications) is heavily reliant on the strength of your marketing strategy. Many people I have met in the property industry miss the important step of having a strategy and are therefore disappointed with the results of their operational marketing.

As many people in the property industry are unfamiliar with what "real", strategic marketing is all about, I thought I would tackle this subject separately.

Furthermore, if you hire a marketing or a PR agency or even take the big step of employing a marketing person in-house, their effectiveness will be reliant on the partners having thought through the following issues and coming up with a strategic plan so that the marketer or PR can then use their skills to decide which tools and techniques they should implement to achieve your aims by following your chosen strategy. Of course, you can hire strategic marketing consultants (which is where people like me come in) but you can save yourself a lot of consultancy fees if you at least start to grapple with these issues yourself.

Why marketing your practice is different to marketing a building

Often, agents will tell me that as their clients pay them vast sums of money to market developments and buildings, they are experts at marketing. Yet if you look at the curriculum for surveying school there is not that much about marketing. Even if an agent has taken the time and trouble to study marketing they are typically focusing on mainstream business to business marketing (B2B — for commercial property) or consumer (B2C — for residential property). Whereas what you are promoting as a firm of surveyors, agents, architects or civil engineers is a professional service. Let's look at some of the main differences.

Table 4.1: Differences between marketing a property and a professional service

Marketing a property	Marketing professional or agency services
Tangible — you can see it, touch it, feel it	Intangible — advice or services you can only experience
One location	Variable locations
Fixed qualities — once built the size, walls, ceilings and windows are fixed	Variable/flexible qualities — the people providing the services can change and be changed depending on the needs
Clear and certain price — within reason you can negotiate	Variable cost — sometimes there is a percentage cost and sometimes an hourly rate, but often it is hard to know how much things will cost
Small target market — often alerting a limited number of other agents	Large target market — there are many individuals and organisations who may need your services
Task is to attract a buyer or lessee	Task is to attract the vendor or lessor
One-off transaction — once it is sold or let, it is over	Ongoing relationship — there are likely to be numerous instructions over a period of time
Task is to build the brand of the development	Task is to build the brand of a firm or team of people

Source: Kim Tasso

In essence, it is to do with the difference between transaction marketing (doing the deal) and relationship marketing (building a profitable relationship). For marketing properties you are often using advertising and PR, whereas in professional services marketing you are mostly using sales techniques (which is covered in chapter 6). The first uses cash, the second uses plenty of time and, for a professional practice, time is money. Yet, while the markets are being as miserable as they are at the time of writing (January 2009), there are many surveyors and agents with time on their hands.

So what is real *marketing?*

Hopefully you already realise that there is more to marketing a professional service than a logo, a brochure and an advert in the property press. But what is real marketing? Let us start by considering the various linked business development processes within a professional practice.

Figure 4.1: Integrated business development processes

Marketing
- Strategic audit, research and analysis
- Service development
- Market development
- Segmentation and targeting
- Pricing
- Brand/profile building

Relationship development
- Retention and development
- Client care and cross-selling
- Relationship management (CRM)
- Key account management (KAM)
- A plan for each major client or referrer

Selling
- Acquisition of new clients
- Enquiry handling/conversion
- Meetings and presentations
- Estimates/quotes
- Pitches/tenders (EOIs/RFPs/ITTs)
- Specific client/needs
- Pipeline management
- A plan for each key prospect

Source: Kim Tasso

The starting point is the business plan (see chapter 2), which sets out what the firm (ie, the partners) is trying to achieve overall. Then, if there has not been sufficiently detailed analysis of the firm's past and present marketing activities and its external competitive environment, then there will need to be some work in this area. This may involve some research — into markets, competitors and clients rather than properties. The firm must then make some tough decisions about which services to promote into which markets (or segments), at what price and with what promotional support.

The headline marketing strategy is then written down into a marketing plan, which is communicated to all those in the team. The purpose of the marketing element is to position your firm and its services in a different place to your competitors, so that it is easier for your target audience to choose you rather than your competitors.

Marketing strategy is about deciding which services to promote into which markets — as it is impossible for most firms to promote all their services into all their markets at the same time with the same vigour. Although as smaller partnerships tend to operate on the basis of "every partner ploughing their own furrow" you often find firms trying to promote every service into every market with the resulting diluted or confused message getting into the market. Or, more commonly, a few energetic partners promoting their services successfully, which can lead to the firm growing and developing in a way that was not quite foreseen at the outset.

Once you have managed to initiate a dialogue with your prospects — and this is the task of marketing — you move from the general needs of those in the market (eg, to move to a new corporate headquarters, to challenge your rating revaluation, to purchase a new home in a particular location) to the specific needs of that particular client and you start selling. Sometimes this is very informal — just meeting and chatting — and sometimes it is very formal — such as when you are sent a big pile of tender documents. Once you have converted their interest into business you have created a client.

The task now shifts to developing a relationship with that client — ensuring that they are happy with what you do for them, introducing other people and services in your firm to them (cross-selling) and strengthening the relationship. You might also have to do a lot of relationship development with intermediaries and those people who refer work to you.

Figure 4.2 illustrates how you can think about the business development process in a different way.

Figure 4.2: The business development process

As you can see there are different comfort zones depending on whether you are a traditional marketing expert, an agent or a surveyor — all we need to do is to get them to work together! This chapter deals with marketing planning. Chapter 5, on tactical marketing, deals with awareness raising and lead generation. Chapter 6, on selling, deals with winning business and chapter 7, on clientology, deals with account management.

Checklist and actions

- What is your present marketing strategy?

- What marketing works well and what do you need to improve?

- Does your current marketing strategy support your business mission and aims?

- How well do you understand the difference between marketing clients' properties and marketing your own practice?

- Assess your firm in terms of its effectiveness in marketing, selling and relationship development. Where are you strong? What needs to be improved?

So what is strategic marketing?

Still confused? Perhaps it might help to think generally of the difference between strategic and tactical decisions. Next time you make a decision, test it against the eight ways to distinguish between strategic and tactical decisions in chapter 3 on leadership in the section on "The Helicopter view — strategic thinking". If it's tactical you should probably delegate it to someone else. If it's really tricky and hurting your brain and your partners are arguing with you, chances are that it's strategic!

Inside out thinking

The job of marketing is to think from the market and client's point of view rather than from the firm's or partners' point of view. In the business plan you have an internal focus — you are thinking about your aims, your strategies, your people, your resources, your technology and buildings and so on. When it comes to marketing, we have to step outside our firms and see things from an external perspective. What is going on in the world and how this might affect our current and future clients and their present and future needs? You can then start thinking properly about how you adapt your firm, its people, its services, its prices and its promotion in order to be well positioned to meet those needs and make a profit.

This is often an alien and difficult challenge to surveyors and agents. But it is the bread and butter of marketing folk.

Only when we have a really good understanding of this external perspective do we consider how it might affect our business — the nature of the opportunities and threats that are out there and how these interact with our strengths and weaknesses. Also how they might impact on what we aim to achieve. Here we have to do something else that doesn't come naturally to property professionals — we have to take a long-term look at things, we have to think laterally and creatively about what economic or market changes might mean to our future clients and we have to think about how we might adapt the way we do things or the services we provide to meet new and perhaps strange new needs.

The marketing audit — strategic analysis and research

Much of what you need to do in a marketing audit should have been covered in the business planning section. But knowing that many in the property industry are hard pressed for time I have summarised the marketing audit process and tools here as well. If you have read chapter 2 on planning (and completed the actions) then well done and you can skip this section.

First you stop thinking about property. Yes, that's right. Imagine that you are an economist or a business analyst. You also need to imagine that you are a fortune teller with a penchant for research. So we start by looking at the big world out there and thinking what might

Figure 4.3: The marketing audit

Source: Kim Tasso

be happening in the areas of economics, social trends, politics and legislation and also technology and researching or thinking about what big changes there might be in five or 10 years' time. We call this a STEP (or PEST if you are in a negative frame of mind) analysis. In chapter 2, there are example exercises to show you what you need to do.

Then you carry out a regional analysis. Again, there are examples in chapter 2. And then a property sector analysis — what is happening in retail, industrial, agricultural and other markets where you are involved. Sometimes, property businesses have an industrial and commercial sector approach — they concentrate on particular sectors, such as care homes, education, hotels, leisure. This requires a further piece of research and analysis.

Business sector analysis

Commercial clients like it when you have a deep understanding of their markets. It means that you can speak their language and it saves them time having to explain the idiosyncrasies of working for, say, a local authority or a hotel or a college of higher education.

Many property businesses will do a sector analysis — but consider only the direct property issues in that sector — how many new starts, vacancy levels, rent rates and such like. However, to be effective business advisers you need to consider the general trends that are taking place in that sector. For example, if you are focusing on the retail sector you will consider all the business trends and issues that are taking place there. You will be familiar with the key media in that sector (eg, *Retail Week, Supermarketing, The Grocer*) and you may even subscribe to some of the main e-alert news services that exist in the sector (eg, *The Retail Bulletin, Verdict retailfreeview*). This way you will be attuned to the issues in the heads of the senior management of retail companies and be able to talk to them on serious commercial issues — and hopefully talk about the future property implications of major trends and changes.

The market–service matrix

Another way to engage with the concept of market sectors is to review your income in terms of what services are promoted into which markets. Marketing people are into matrices. The matrix is a vital business analysis tool — and you could spend some time looking through this book to count how many times we use a matrix.

But I digress. This market-service matrix can often be very illuminating — showing which services are, for example, the door openers — the ones with which it is easiest to gain a foothold with a client. There may be some silo services too — where it is relatively easy to promote the service to a client but rather difficult to cross-sell other services after this. It might also show you which services it is easiest to sell to clients in all sectors. We will discuss cross-selling to existing clients more in chapter 7 on clientology. For now, we are concentrating on identifying trends in the way in which you promote and deliver different services into different markets and where there are significant gaps for you to fruitfully explore. Often, this analysis will reveal that a market-facing approach has more strength in strategic terms than promoting individual services to a wide audience.

In figure 4.4, we can see that the practice concerned is really strong in the retail sector — all of its services are promoted there — whereas in the finance sector it is only effective at selling valuations, investment agency and management. It may be that the marketing strategy — which is about deciding which services to promote into which markets — decides as a result of this analysis that it concentrates its strategy on further active development in just the retail and house builders sectors.

Figure 4.4: Markets and service matrix

Market sectors ➤

	Retail	House builders	Schools	Finance providers
Valuations	Yes	Yes		Yes
Investment agency	Yes			Yes
Building consultancy	Yes	Yes	Yes	
Planning	Yes	Yes	Yes	
Management	Yes		Yes	Yes
Industrial agency	Yes			
Rating	Yes	Yes	Yes	

Services/expertise ↓

Source: Kim Tasso

Boo hiss — competitor analysis (revisited)

You also need to take a long, hard look at your competitors. Hopefully you did this in chapter 2. You didn't? It is important, so I will try it from a different angle.

Let us do some research into our competitors. This does not mean the nearby property firms, or looking at their tactical marketing efforts (eg, their brochures and where they are advertising). You need to look at a broader range of potential competitors where clients are obtaining the type of advice and services for property-related matters where they should be coming to your firm.

Start with your obvious competitors. Develop a table that shows the critical information about your competitors and compare this to your firm. Consider the number of professional staff that they have, the services they provide, their key clients, the research they produce and offer to clients, their pricing, their online services, their growth rates, their management structure, their key people and so on. The things that you need to consider will vary depending on a number of factors — such as your key services, the areas and types of client you serve and how you position yourself in the market (eg, low-cost, high-volume services for consumers or high-cost, strategic services for major companies).

This sort of analysis can easily be allocated to junior members of your team. Explain what you want them to do and let them loose on industry associations, trade media and the internet. Give them access to online research resources (eg, OneSource, Experian) so that they can profile the companies, firms and individuals at your competitors.

Once you have analysed the obvious competitors, consider the less obvious ones. What are banks, solicitors, accountants, consultants and other specialists doing to nibble away at your relationships? What can your clients and potential clients do more easily online using cyber-intermediaries and other portals? To what extent can clients obtain the relevant property services "bundled up" with other professional or financial services? What about outsourcing?

Take some time to consider all the existing and potential alternative sources of advice that your clients can obtain, and think about how your firm's services stack up against them. If your clients really can obtain a similar or better service at a much cheaper price online you can be sure that you are going to lose this business in the short to medium term and need to do something about it rather than hide your head in the sand. It is scary, I know, for I have been there. I have watched people who are far less qualified, less experienced and less able than me at certain jobs put together a great package that offers a similar service at a much lower price and promote it effectively.

Put yourself in the client's shoes and consider why you might choose your firm over the wide range of competing offers. Be brave and do the analysis — you can only start to formulate strategic solutions if you confront the inevitable at the earliest possible opportunity. Some people find it easier to produce their competitor analysis using a framework such as Michael Porter's Five Forces, which is shown in the chapter 2.

Some firms have found it easier to get lots of people involved in competitor analysis — either by allocating particular competitors to specific teams to investigate different competitors or by "nominating" one competitor each week or month and getting everyone to submit their knowledge and information about that competitor to a central person for compilation. Often in the property industry, most of the useful competitor information will be in the heads of your people who are out there in the market talking to clients and intermediaries and observing, first hand, the actions or results of competitors. Finding a way to collect, record and manage such sources of information is difficult but it is worthwhile. Of course, if you have constant monitoring systems in place — and an easily accessible central source of information — such as a competitor area on your intranet, then it will become second nature for people to collect and share such information on an ongoing basis.

Another fruitful source of competitor information is from clients. Many clients will use more than one property adviser and some may be willing to share their perceptions of the differences between firms. If you are required to submit a tender for a major piece of work you may find that the client will provide you with a detailed debrief on how you compared against your competitors. Larger clients — such as public bodies — will have automatic debriefing processes where the scores for your firm are shown against the highest and lowest of all those tendering. This provides a useful test for how strong your particular team and offer was on one such occasion. But you need to analyse a wide range of competitor sources before making judgments about your firm's strengths and weaknesses generally.

If you end up collecting a lot of information, you may need a way to organise and manage it succinctly. Some firms set up a document or proforma to organise all the relevant information about competitors in a similar way — along with a "recent developments" section so that new information can be added as it is obtained. Some supplement these "one-sheet competitor profiles" with a section at the end that says something similar to "What this means for us when we are competing against them" — so that you can summarise your competitive response to whatever it is you see as their strengths and weaknesses and differences.

Some firms decide to summarise on a sort of "dashboard", where they use the collected data to make subjective judgments and assign a rating to a number of different factors or aspects of that competitor (see figure 4.5).

Figure 4.5: Assessing your competitive position

Competitor	Profile	Size of specialist team	Financial performance	Growth	Office network	Use of technology	Total
Our firm	2	8	9	7	5	2	33
Firm A	10	6	2	4	3	7	32
Firm B	6	2	6	9	2	9	34

Source: Kim Tasso

Checklist and actions

- Decide how you will tackle your marketing audit. Who will help you? How much time do you need? When do you hope to complete it?

- Review the analysis you completed as part of the business planning chapter.

- Complete a STEP (or PEST) analysis and discuss it with your colleagues.

- Ensure that someone (or somebodies) complete the relevant business sector analyses.

- Produce a market-service matrix to look at your practice from a helicopter perspective.

- Allocate responsibility to partners or teams to complete the various competitor analyses that you need. Ensure that you put in place the relevant systems to continue to update and review competitor information on a regular basis.

The crown jewels — key client analysis

Now you need to really get to grips with information about your past and current clients. Some ideas for analysing your clients are shown in the chapter on clientology so have a quick look at that section now and aim to produce the following for the marketing audit.

For those in the commercial space, look at your top 100 or so clients. Pareto indicated that 80% of your business will come from just 20% of your clients — so these clients will represent the lion's share of your business.

Build a chart with their overall fees generated over the past three years. Look to see whether they are mostly increasing, mostly decreasing or pretty stable. Then look at the breakdown of fees by different service areas — for example, by valuation, by investment agency, by building consultancy and by planning. See where there are the most gaps, and then build profiles of the types of clients that they are. For example, what sector are they in, how large are they, are they mostly private concerns or public companies, where are they based and how did they get in contact with you in the first place?

An example of this sort of analysis is shown in table 4.2.

Table 4.2: Major client analysis and cross-selling matrix

Last updated (Date)

CROSS SELLING MATRIX ANALYSIS 2007

Client name	Partner	Sector - Business	2005	2006	2007	YTD 2008	Projected 2008	Agency	Valuation	Management	Investment	Building Consultancy	Notes
Accountancy heroes	B	Professions	111	222	333	155	333		333	333			International potential
Airline One	C	Transport	111	111	111	111	111				111		
Association	X	Charity	111	222	222	111	222	222					Overseas potential
Bank 2	D	Bank	111	222	444	333	555	222			222		
Big firm	Y	Professions - Property	888	999	777	333	666	666					111
Buy me	Z	Retail	0	999	0	111	555					444	
Connections Plc	T	Telecoms	666	555	444	111	333						Going in house?
Consulting 2	X	Business services - Marketing	999	222	888	888	999	888				111	
Data company	G	Technology	555	555	999	888	999				888	111	
Dealership	G	Distributor - Car dealership	111	222	111	222	333		222				
Designers Inc	B	Importers	111	111	222	222	222						At risk
Developer 4	Z	Property - Development	111	333	0	111	444						111
Eat well and be merry	Z	Retail	0	0	888	555	555	777					
Finance house 1	G	Finance	222	333	444	333	555	444	111		333		Financing work
Fund	B	Charity	444	444	444	444	444						
Get together today	F	Charity	111	111	0	111	111						
Give well	D	Charity	666	666	0	0	0						
Government	X	Government	888	888	999	777	888	999					Active research programme
Heating Contractors Ltd	B	Construction	222	111	222	111	555		222				
Help them	F	Charity	999	888	777	333	666						
High Street shop	F	Retail	0	111	222	333	555	777			222		
House builders plc	Y	Property - Housebuilders	888	222	444	222	666	444					
Household stuff	T	Distributor	222	222	0	333	666				333		Referred by xxx
Insolvency firm	F	Profession	0	111	333	222	555						
Lovely property company	X	Property	0	0	999	222	444	999					
Manufacturer 2	D	Manufacturing	999	888	777	333	666		444			777	
Media Group 3	B	Media - Publishing	555	666	444	0	0						Pitched and lost
Mega Marketing company	X	Business services - Marketing	888	888	888	888	888	888				222	
Partners United	G	Professions - Property	333	222	333	111	333	888					111
Property Co 4	Y	Property	222	666	888	444	999	888			333		
Publish and be damned	Y	Media - Publishing	111	555	333	222	444	444					
School	D	Charity - Education	111	111	111	111	222	111					
Secondary foreign bank	G	Bank	444	555	444	222	444		444				On hold for present
Shop till you drop	A	Retail	888	0	0	444	444						
Small charity 4	Y	Charity	333	555	555	111	444	555					
Systems 1	D	Technology	0	444	333	333	222				333		
Wired for sound	D	Telecoms	0	0	0	999	999						New for 2008!
Total			13,320	14,319	15,429	11,588	19,203	8,880	1,776	333	2,775	1,665	15,429
Growth rate				8	8	-26	66						

Source: Kim Tasso

Of course, as well as looking at the numbers in terms of revenue and, more importantly, the profit from each client you should also look at their satisfaction and their propensity to recommend you. This can be assessed with a simple client research exercise, which is examined in more detail in chapter 7 on clientology. As well as assessing the levels of satisfaction, client research interviews are also a valuable source of information for your marketing strategy. Clients will often tell you why they choose your firm, the strengths and weaknesses of your firm against your competitors and will provide insight into your present reputation and how you need to improve both your marketing and your relationship management. Sometimes, client research will also reveal clues as to how you develop your services and markets that will form a fundamental part of your marketing strategy.

More importantly, if your practice is one that relies mostly on a few critical client relationships then it is pointless to think about a marketing strategy that reaches hundreds or thousands of potential clients. With a business that relies on just a few really important client relationships, your marketing strategy may focus almost entirely on programmes that are designed to strengthen and develop those existing relationships and your new business development strategy is honed down to identifying just a handful of prospects that you need to nurture over time. This is more of a sales strategy than a marketing strategy but it is just as valid. Also, it will save you time discussing inappropriate marketing communications and profile-raising ideas. If this sounds similar to your firm, have a look on the section below about "Dream clients" and then turn to chapter 7 on clientology.

Where does it come from? Source analysis

For those with mostly consumer clients (ie, those in the residential market) you need to analyse where these clients came from. Do you have the right systems to track the source of each case or assignment? No doubt a proportion of clients will come from your marketing activities — but do you know which ones? And many will come from existing client referrals or from referrals from other intermediaries but do you know how many and which ones? You need to be able to break down the different types of new clients or assignments from the various sources in order to focus on those that are most worth development as part of your marketing strategy. One firm I worked with took almost two years to develop the various systems (new forms, changes to accounting systems, training of staff to record the data, procedures for entering and verifying information, management reports etc) to enable them to accurately analyse exactly where the new instructions (in both volume and value terms) came from. The results they obtained enabled them to make informed decisions about where to invest their limited marketing resources in the future.

Table 4.3: Client and work source analysis — transactions

	2006	2007	2008
Existing clients			
Known	24	58	45
Unknown	17	26	19
Referrers			
Finance org 1	28	28	19
Finance org 2	2	18	17
Finance org 3	0	0	67
Other financial	57	67	125
Solicitor 1	16	18	16
Solicitor 2	5	8	8
Other solicitors	4	2	2
Other referrers	19	22	24
Events			
County show 1	28	29	30
County show 2	12	13	12
School fete 1	4	5	8
Business trade show	2	10	3
Promotion			
Adverts — local	3	13	15
Adverts — property	2	3	2
Adverts — directory	1	4	5
Online advice site	0	3	29
Online property portal	18	14	2
Direct			
Mailer 1	29	69	39
Website	15	33	58
Unknown	67	87	114
Total	**353**	**530**	**659**
Growth		**50.14**	**24.34**

Source: Kim Tasso

While this is an illustrative example using anonymous data, you can see how useful it might be to you in developing your marketing strategy. For example, it shows you that your events are not generating much work (although they may be contributing hugely to your reputation within the community so think about other ways to measure success — such as through interviewing clients to see whether they regard your presence at such events as important for relationship management or your local profile). It also shows that there is a lot of work where you do not know the source and need to investigate further and/or change your data capture systems. Furthermore, it shows the huge importance of referrer relationships in your marketing. This would suggest that you can stop wasting time and money on some of the more obvious promotional methods and start investing more time and effort into a structured programme of relationship management. This would be a great strategic decision.

Checklist and actions

- Speak to accounts and other support staff so that they produce the required key client analysis for the past three years.

- Ask them to produce a cross-selling analysis of those same key clients for the present or previous year.

- Check that you have the systems, procedures and behaviours in place to track the source of new clients and new instructions.

- Discuss these analyses with your partners — on a regular basis.

- Consider organising some client satisfaction surveys.

Internal analysis

As part of your business planning exercise (see chapter 2) hopefully you will have a good grasp of all of the relevant financial, human, service and other internal information.

For example, you should know the income and profit profile for the last few years from each of your different divisions, offices, service or market teams. You should be aware of what parts of your business are growing, which are in decline and which are stagnant. You should be aware of the profitability of the various parts of your business. However, I know that with some firms, who do not monitor their recorded time, calculating profitability can be very tricky. But you can do a similar exercise with revenue and have a go at calculating profit by taking a small sample of instructions and getting someone to guess and manually calculate profitability.

You should also know what time is invested by different partners and fee-earners in marketing and business development, as well as their "production" time actually working on client matters. You should understand the extent to which the money set aside for marketing and client entertaining has been spent on firm-wide matters (eg, for brochures, logo, website, firm events), as opposed to particular divisions or teams or services.

Again, there are a vast number of areas for you to consider but the strength of your marketing strategy will depend on the strength of your analysis and understanding of your market and how it is likely to change in the future.

Developing a marketing strategy is essentially about making choices. So make sure that you are equipped with the necessary information that will support good choices. And avoid unnecessary rows between partners — you can be sure that your partners will challenge your suggestions and they will start by looking at the quality of the information that you use to build those ideas.

Stars, dogs and cows — portfolio analysis

An important part of your internal analysis will be to consider all the products or services within your portfolio. This is an area of marketing analysis that forces you to look beyond individual partners, offices or services to consider the overall mix of income in your business.

Many markets and services have what we call a life cycle. It means that they take a while to get going, then you spend some time enjoying fast growth and increasing profits, and then either demand declines or competitors pile into the market with cheaper/better alternatives and then many — without strong action — will go into decline.

Figure 4.5 illustrates the life cycle.

Figure 4.5: Product life cycle

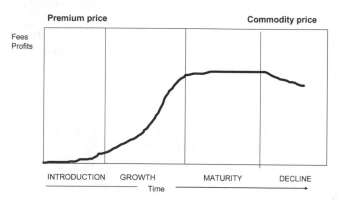

Source: Theodore Levitt, Harvard Business Review (1965)

An example might be rating revaluations. In the run up to the next revaluation, the smart firms will start mailing and talking to their corporate clients well in advance of the revaluation and securing a number of small preparatory jobs. As the revaluation date grows nearer, more of your competitors will be having those same conversations with their clients and also with your clients — who are their prospects. The market will accelerate and the work pace will quicken. Everyone tiptoes around the rating team as they are stressed, stressed, stressed. Then the revaluation date arrives and there is some work on appeals and such like — but now everyone is offering rating support and the market has tired of hearing about it. Meanwhile, the amount of work declines to the normal level. The only way to win extra work and sustain the fast growth of the past months is to market more aggressively or do something really neat with your pricing. Now think about what the appropriate marketing strategy would be at each stage of the cycle — where do you think that your bucks will generate the biggest bang? How should your marketing change through the cycle? Do you want to be a leader or a follower? Is rating going to be one of your major campaigns or will you focus on another area where there is either greater potential for ongoing work or where there is less competition or where you have a stronger and larger team?

The life cycle idea is important because it has many implications for aspects of your marketing strategy — particularly the price you might charge and the appropriate marketing communications approaches required.

What is perhaps more important is to consider all the different products within your practice as no doubt there will be some that are at the start of their life cycle and some that are in the maturity or decline phase. You need to consider all the products/services at your firm to gain an understanding of the overall picture. This is where another core marketing concept comes in — the idea of a portfolio. And here we need the matrix to help us consider the situation. There are many sophisticated tools to look at this but one of the simplest and most robust is called the Boston Consulting Group tool.

Figure 4.6: Boston grid matrix/portfolio analysis

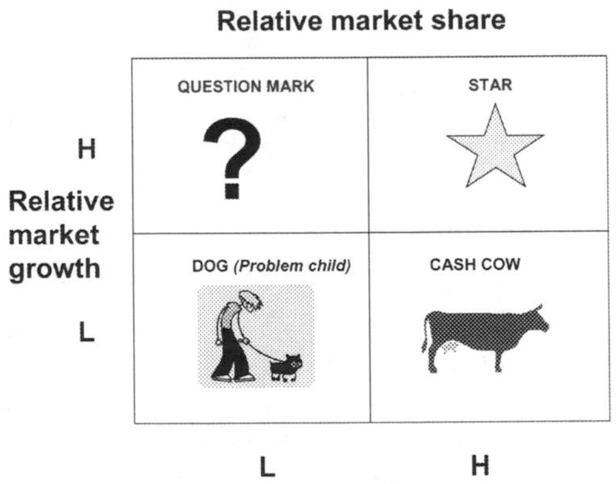

Source: Boston Consulting Group

What this analysis does is help you pinpoint where you are generating the majority of your cash and profits today and where you might generate these from in future years. Of course, the danger with many firms is that they assume that those products or services that produce the majority of income today will always do so and therefore fail to invest sufficient time and money in ensuring that you convert enough dogs and question marks into tomorrow's stars and ensuring that today's stars become tomorrow's cash cows. It may be stating the obvious, but your investment and development strategies for your various products and services will be a key element of your marketing strategy.

Summarise with a SWOT

You have now spent some time collecting information and immersing yourself and your colleagues in that information. You have decided which — of all the various ideas and issues you have identified — are the most important. You have maybe even argued with your partners about what are the important things to address within your strategy — and what the implications are for your practice.

As with the business planning situation (see chapter 2), you need to review all the information that you have gathered during your marketing audit and focus down on the most important issues for your firm and its various markets and services. It may be that you allocate the various tasks to different teams or partners in your practice. But at some point you need to come together and look at the key findings and agree on what issues you will concentrate your efforts.

By summarising the key strengths and weaknesses and how these translate into opportunities and threats you should see some patterns or key issues emerging. If not, then go back to all the piles of analysis and information and start trying to weight the different issues in terms of their importance and/or potential impact.

Checklist and actions

- Review the internal analyses on the financial performance of each office, team, service, market or partner that you completed as part of your work on chapter 2 (business planning).

- Organise any additional financial or internal analysis that you need.

- Review the non-chargeable time spent by your partners and fee-earners on marketing, selling and relationship management.

- Complete a portfolio analysis and think about the product/service life cycle at your firm.

- Summarise all the analyses you have completed with a SWOT.

- Identify the critical and key issues that you need to address as part of your strategy and discuss them with your partners.

Setting SMART objectives

Hopefully, the data collection exercise will have provided you, with among other things, some benchmarks that show you your current position. This will enable you to set SMART (Specific. Measurable. Achievable. Realistic. Time specific) objectives for what you wish to achieve with your marketing strategy.

In the earlier business planning section, we looked at SMART objectives for the firm in terms of the overall income and profit and other firm-wide aims that the firm wishes to achieve. Marketing objectives are slightly different, as they are focused primarily on what income and profits you wish to receive from particular services and markets. So, for example, your marketing objectives might look like this:

- Generate 10% additional revenue from market X within two years.
- Increase profit from service A by 15% by December 2010.
- Introduce two new services to market X within three years.
- Gain 10 more clients generating over £5,000 a year in market Y within 18 months.
- Introduce service B to 10 new clients in market Z by July 2010.
- Win five new clients for service C with an income of at least £3,000 pa by December 2010.
- Gain a further 10% of market H by March 2011.
- Change the balance of the business to being: 33% service A, 33% service B and 33% service C within three years.

When you are setting objectives — and, more importantly, getting the rest of the partnership to agree to them — you will find that you have to make some hard choices. For example, if you want to grow one part of the business faster than another, then you will need to devote more resources to the part you want to grow faster. And if you are keen to build your profile and reputation in one market for a particular area of expertise then it may mean that your reputation for something that you currently do is likely to diminish. Of course, if your overriding objective is to maximise profit in the short term, then this will conflict directly with your ability to invest and generate profits for the longer term.

Strategic choices — competitive advantage, target markets and priority services

What competitive advantage and scope?

Michael Porter, one of the great business gurus, articulated that there are very few choices when it comes to deciding how you will compete. In essence, he said you can either do it by being the cost leader (ie, the cheapest — so a pricing strategy and/or an automated product strategy — see below) or you can do it by differentiation (ie, a branding strategy — again, see below). He also indicated that the main choices you have on market scope is to try to tackle a broad, mass market or you can select just a particular part. He had a matrix to show this.

Figure 4.7: Competitive advantage and scope choices

Competitive scope

		Narrow	Broad
Competitive advantage	Differentiation	Differentiation Focus	Differentiation
	Low cost	Cost focus	Cost leadership

Source: Michael Porter

The professions are littered with sad examples where firms get this completely wrong. The most common situation is where a firm tries to promote a "Rolls Royce" service but at a "bargain basement" price. Thus they fail to make any profit and go bust. If you work hard at developing a service or proposition that is truly different from your competitors' — and in a way that truly adds value to your clients — then please do not be tempted to offer it at a cheap price. Partly because if you offer something special and different at a really low price then you undermine its credibility and you devalue what it is that you are offering. If you really do have a commodity product only — then, by all means, provide it at the lowest possible price. But do not mix the two strategies. The easiest way to think about it is whether you are trying to maximise the volume or the value of the work that you do.

There is much in the business books about competitive advantage. Treacy and Wiersema suggested that there are three "dimensions of value" or strategic positions:

- **Product leader**
  Technology frontiers — this means that you have invested in technology or special processes and are recognised as a leader in a particular area of expertise. Many firms at the forefront of internet property sales succeeded this way.

- **Operationally efficient operator**
  Reliable and dependable — this means that you have found a way of working that is faster or more efficient and cheaper than others. For example, you may be using less qualified staff or have operations in a low overhead environment.

- **Client intimate firm**
  Responsive and flexible to match precise needs of different clients, some firms are known to be closer to their clients — often through long-term relationships and close collaboration — than others.

They argued that there were four rules of success:

- Become best at one of the three dimensions of value.
- Achieve adequate performance in the other two.
- Keep improving one's superior position in the chosen area so as not to lose out to a competitor.
- Keep becoming more adequate in the other two areas, since competitors keep raising client expectations about what is adequate.

Hugh Davidson wrote a number of books — primarily aimed at consumer markets — and he identified the following eight most significant competitive advantages.

Davidson's eight most significant competitive advantages:

1. Superior product benefits (eg, an online system that provides easy access to your online property terrier as well as special reporting tools).

2. Perceived advantage (eg, a high-value brand name, such as Savills, has currency in some markets).

3. Low-cost operations (eg, some agents operate solely on the internet and offer cheaper services than those with buildings, teams of support staff etc).

4. Legal advantage (eg, patents, copyrights, exclusive arrangements with others in the markets).

5. Superior contacts (eg, having an exclusive relationship with a key market intermediary or being well connected in a particular region or market).

6. Superior knowledge (eg, better research or information and knowledge systems).

7. Scale advantages (eg, the largest operators often manage to do things more cheaply than others through investment programmes. They spread their overheads better).

8. Offensive attitudes (eg, Some firms, such as Foxtons, are perceived to be aggressive and market changing).

If these ideas appear somewhat remote from the world of property, then let us look at the thoughts of one of the greatest management gurus ever to focus on professional service firms. David Maister is one of my heroes — make him one of yours too. He also identified eight competitive advantages (set out below). I used these criteria with the equity partners of a large property firm — they all had to rank the order in which they thought their firm competed most effectively. Through discussion we managed to work out the two most critical issues for the firm and to design a strategy to help them leverage those points.

Maister's eight sources of competitive advantage for a professional service firm:

1. Innovative hiring practices.
2. Training excellence.
3. Unique problem-solving methodologies.
4. Special client counselling skills.
5. A particularly robust or valuable knowledge base.
6. A different approach to project and team organisation.
7. Research and development (R&D) excellence.
8. Marketing and/or client listening skills.

Whether or not you use these ideas and frameworks to help you identify your source of competitive advantage you need to know what it is. So, what is it?

Target markets and priority services

A good strategy will have balance between generating profits for today and ensuring there are new profit streams developed for the future. In marketing, we use something called the Ansoff matrix to make these choices clearer.

Figure 4.8: Marketing strategy choices with an Ansoff matrix

Services/products

		Existing	New
Markets	**Existing**	Penetration 65%	Service development 30%
	New	Market extension 45%	Diversification 15%

Source: Ansoff

What this matrix conveys is that there are only four choices when it comes to marketing strategy. The first is to do more of the same — the penetration strategy — selling more of your existing services to your existing markets. This is the most efficient strategy for generating more profits in the short term. However, for generating profits in the future you can either invest in developing new markets — such as a different region or sector — or you can develop new services to promote into your existing markets and clients.

In an ideal world you will have an element of your strategy focusing on penetration and at least some energy to developing either new markets or new services. We leave the diversification area — where we try to do new things in new markets — to those who are either very brave or very foolish.

An extended version of this matrix — to help drive strategic decisions about the choice of markets — is shown in figure 4.9.

Figure 4.9: Selecting markets by matching strengths to attractiveness

		Business strength		
		Strong	Moderate	Weak
Market attractiveness	High	Extend position	Invest to build	Build cautiously
	Medium	Build selectively	Invest selectively	Limit expansion
	Low	Protect and refocus	Harvest	Divest

Source: Kotler

What figure 4.9 shows is that you need some structure to aid your review of the different market options — whether these are geographic markets, commercial business sectors or particular business units. You consider how attractive each market is — perhaps by devising some rating scale that takes into account your overall aims and strategy (eg, growth rates, profitable work, interesting work, prestigious clients, international potential etc) and then consider how strong your view is in each market. Obviously, you will have greater strengths in those markets where you have an established presence or track record so you need to find a way of grading your strength — for example, how many experts you have in that area, how many major clients, how many years' experience or the unique knowledge and experience your firm has over your competitors.

When we look at what the experts in professional service firm marketing have to say, they indicate that there are a limited number of options when it comes to thinking about your market strategy:

Mayson's 10 options for market strategy choice:

1. Market sector or services
   a. focused firm;
   b. portfolio firm;
   c. general firm.
2. Domestic geography
   a. local firm;
   b. regional firm;
   c. national firm.
3. Foreign geography
   a. Four choices.

In essence, Mayson is simply setting out the obvious choices that you have. You can either focus on just one area or spread yourself across a number of areas or even provide services in all markets and all sectors. Then you consider your geographical spread — local, regional or national. Then some firms will have international aspirations and here you can have a loose association with an overseas operator, an exclusive or formal relationship, a merger or even send out your own people to set up an international office from scratch.

Finding your niche — segmentation strategies

Sometimes, and particularly if you are a small firm with limited resources and/or little appetite for marketing, you may decide to select just one small market and make it your own.

As part of your business planning and/or marketing strategy exercise, you need to consider the various markets you serve and how these markets might be broken down into distinct submarkets. There are numerous ways to do this and obviously the approach you adopt will vary depending on whether you are focusing on commercial markets or on consumer markets. In broad terms, the way you segment your market is shown in figure 4.10.

Figure 4.10: The market segmentation process

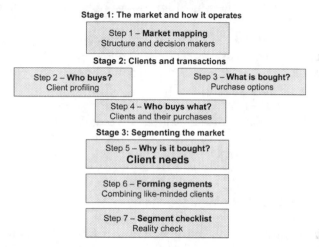

Source: Malcolm McDonald and Ian Dunbar *Market segmentation. How to do it. How to profit from it.*

To illustrate the point, you could consider one of the following common segmentation strategies:

- Commercial
    - By geography (local, regional, national, international).
    - By type of property (retail, industrial, commercial etc).
    - By market sector (retail, technology, agricultural, telecoms, schools etc).
    - By size of organisation (revenue or capitalisation value).

- By legal structure (sole traders, partnership, private company, public company).
- By growth rate (start-up, fast growth, mature, declining).
- By issue (automation, international competition, downsizing, merger).

- Consumer
  - Location (local, regional, national).
  - Demographic (age, socio-economic group, race etc).
  - Life cycle position (single, married, young kids, empty nesters).
  - Wealth (low, medium, high, super-high income or asset level).

Of course, you need the data in order to segment your existing clients and any prospect lists you might have compiled or purchased. Sophisticated organisations use commercial segmentation tools also — to look for consumers with particular profiles in specific locations. A well-established segmentation and targeting tool is ACORN (A Classification of Residential Neighbourhoods — produced by CACI). This takes into account a number of factors and enables you to see the profile of a particular area — right down to the lowest postcode level of just six households. The top level classifications are shown in table 4.4, although each category breaks down into more detailed subgroups.

The point of all these thoughts about how you break down your market is to identify one or two specific segments where you have some competitive advantage (eg, a number of existing clients who can introduce or vouch for you, specialist knowledge of the particular challenges and requirements faced by those in the sector, particular services that are packaged to meet the special needs of that segment etc). As a segment is usually small and well defined, it means that you can concentrate your marketing efforts and have a bigger impact than if you were dabbling in a larger, broader market where there are larger competitors. It means that you can tailor your service, proposition, message and marketing efforts in a more focused way.

Guard against segmenting your markets based on your products and services — this is too inward looking. The test of a good segmentation strategy is as follows:

Testing your segmentation strategy
1. Is the segment homogenous?
2. Is the segment measurable?

Table 4.4: ACORN classifications

Category	% UK population	Group	Description	% UK population
1 Wealthy achievers	25.4	A	Wealthy executives	8.6
		B	Affluent greys	7.9
		C	Flourishing families	9.0
2 Urban prosperity	11.5	D	Prosperous professionals	2.1
		E	Educated urbanites	5.5
		F	Aspiring singles	3.8
3 Comfortably off	27.4	G	Starting out	3.1
		H	Secure families	15.5
		I	Settled suburbia	6.1
		J	Prudent pensioners	2.7
4 Moderate means	13.8	K	Asian communities	1.5
		L	Post industrial families	4.7
		M	Blue collar roots	7.5
5 Hard pressed	21.2	N	Struggling families	13.3
		O	Burdened singles	4.2
		P	High rise hardship	1.6
		Q	Inner city adversity	2.1

Source: CACI

3. Is the segment accessible?
4. Is the segment substantial?
5. Is the segment exclusive?
6. Is the segment stable?
7. Is the segment recognised by the customers themselves?
8. Is the segment recognised by the intermediaries?
9. Will the products or services be premium priced?
10. Will the segment offer above-average returns?

While having a clear understanding of your segmentation strategy is important for all organisations, those who have decided on a niche strategy will then have to build their entire marketing programme (or campaigns) around a particular niche.

Dreaming of your perfect client?

If you are a small organisation or if your practice relies on a few critical client relationships then it is inappropriate to have detailed segmentation strategies. Your approach might be to do some detailed research on the sorts of organisations that would be your dream clients and then to identify a handful of target organisations. This makes it easy to target action (which is typically more sales-driven than marketing) on those particular organisations. However, you will need to ensure that you have a structured sales programme in place as your main strategy — and this is covered in chapter 6 on selling.

Checklist and actions

- Set some SMART marketing objectives — and ensure that they support your overall business/firm objectives.

- Decide on your competitive advantage and scope.

- Segment your market and identify any relevant niches.

- Check that your segmentation approach is robust and reliable.

- Agree your target markets (segments) and priority services.

- If in a commercial market, develop a profile of your dream clients and try to identify the best prospects.

Innovation and new service development

If your analyses (particularly the portfolio and competitor analyses mentioned above) indicate that your are under increasing price pressure because you are offering a commodity product then your strategy may require you to do something more drastic, such as package (or bundle up) a particular service or to design a completely new service.

The easiest thing to do here is where you find that you have provided a particular service to a number of different clients already. You can "package up" your approach to make it more tangible. Tom Peters, a management guru who spent some time focusing on professional service firms, offered the following guidelines for developing a methodology.

Tom Peters — developing a methodology:

*   How do we define a problem?
*   How do we state the aims and parameters of a particular case?
*   How do we work together with clients?
*   How do we conduct research on the problem?
*   How do we present our findings?
*   What is the course of the problem-solving exercise?
*   How do we inject originality into the process?
*   How do we test our conclusions?
*   What is our definition of quality control?
*   How do we frame the implementation process?
*   How do we distinguish our approach from others?

Management consultants use this all the time to turn "something neat we did for one client" into a "service that all large companies need to get them through an economic crisis". I have used this successfully to lead a team workshop at a property firm where we looked at a particular service area, documented what we did as a team in different situations, thus capturing our best practice from several leading experts and then adding some ideas from some client feedback research we had conducted. A little further work — and some help from the IT team in providing the necessary technology support — enabled us to package the material together with a small, targeted campaign and become the national leaders in a particular area of property finance work. The more defined approach was welcomed by clients as it increased transparency over what was being done, helped the firm deliver a consistent high-quality service each time and provided the necessary differentiating messages for an effective marketing and sales campaign.

If you would prefer a diagram, then you can consider the following service design process, which takes you through a simpler set of processes (Figure 4.11).

You may want some more specific examples of how this might work for your firm. Well, I have recently seen some interesting new "products" from property companies doing things like "disability discrimination compliance check", "Health and safety risk assessments", "Enfranchisement decision support". It is the sort of service that many property firms provide — but it is packaged up in a way that makes it more accessible to clients and easier for you to market to particular groups of clients.

Figure 4.11: Service design process

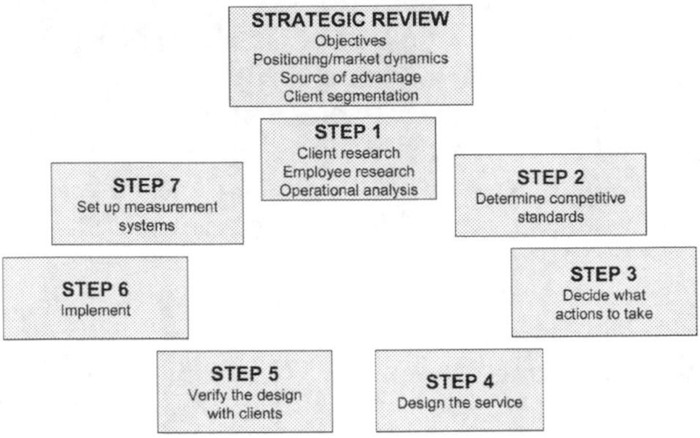

Source: Young and Stone

Pricing

Clients will pay a premium for a brand, and branding is a key marketing strategy for some firms. But you may take a different approach and decide that you can provide the service at a much cheaper level than most competitors and that this is the foundation of your strategy.

The first thing you need to do is to check your numbers and be sure that you can still make the required level of profit if you cut your price (and margins) considerably. Spend some time with a qualified accountant.

Then you need to ensure that you have a carefully packaged "service" that makes it clear exactly what is and what is not included in the price. You should also think carefully about how your existing clients might feel if they see that you are promoting a service at a much cheaper level than they are paying!

You should also be aware that if you offer a low price, then it is more than likely that your competitors will either try to match or better your price — so you have to think ahead and consider how you will respond if this happens. Often low price strategies will only work for a really short time, so you have to ensure that your promotional and sales strategies are suitably geared up to generate a lot of demand very quickly.

Do I need to mention the hourly rates problem? I do? But don't you know this already? You like hourly rates because it helps you with your cost analysis (because you do record all your time don't you so that you can calculate what services and clients are profitable?). And in case the agents among you are smiling smugly because you charge on a percentage fee and therefore do not need to record time, please think again — how do you know what is profitable or not? But we know that clients do not like hourly rates. It is like getting on a plane and the airline telling you that the ticket cost is dependent on the time flying, the prevailing weather conditions and the availability of landing slots at the other end. Clients need certainty. To provide certainty for clients you need to be good at estimating (the quantity surveyors and construction professionals among you are really good at this) what you think the job will cost, the various risks involved and how much margin you are prepared to build in to your price. Fee rates often equate to cost, prices equate to value (to the client). Another reason why hourly rates do not work — clients get irritated if they have to pay the same amount for an hour of your partner's time dealing with administration as they do for a really worthwhile telephone conversation providing strategic input. Furthermore, hourly rate comparisons mean that often the fastest and best advisers end up making less money than the slow, inefficient ones.

Spot the difference — commodities and brands

We have not left the issue of pricing yet.

There are more than 100,000 surveyors in the UK, and tens of thousands more estate agents. What they do in terms of their technical expertise and the knowledge they have is similar. In effect, what they provide is a commodity service. As with all commodities, clients see them as much the same and therefore often choose on the basis of price. Therefore, some firms — rather than expose themselves to constant price pressure — decide to develop a brand for their firm or some of their services. In a professional service firm such as a property consultancy or an estate agency, people may find it easier to think about brand as being like reputation. But brands are about perception — it may be very much the same service — it is just that people perceive (and value) some difference. Think about the branded goods that you buy — are they really any better than the unbranded goods — or do

Figure 4.12: From consultancy to commodity services

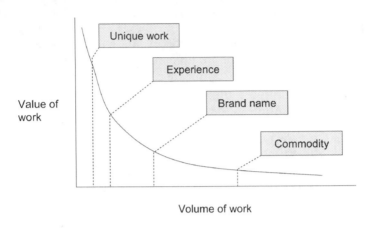

you just "feel" more comfortable with a name that you know and trust or with one that others recognise as having some important attributes?

In case any of you are wondering why I have become "tactical" in the strategic marketing section, a brand is rather different to an identity. They are, of course, linked. But the identity is just one expression of a brand. The identity is the way an organisation is recalled, its association and promise. It is an expression of difference and the sum total of all the differentiating factors we try to create and sustain. Ultimately, it is a firm's only uncopiable feature. The features of a strong identity include: clarity (it needs to be understood internally and externally), relevance (the personality, positioning and philosophy must be meaningful), desirability (in terms of client value), credibility (is it believable? For example, can you really be an international player with just one office in Norwich?) and consistency (keep the message simple and repeat it regularly — stop trying to be too creative at every opportunity — reinforce the key messages, do not reinvent yourself each time).

Definition of a brand: "A successful brand is an identifiable product, service, person or place augmented in such a way that the buyer or user perceives relevant, unique added values which match their needs most closely. Furthermore, its success results from being able to sustain these added values in the face of competition." (de Chernatory and McDonald)

Brands are in their infancy in the property sector. I first wrote about brands in *Estates Gazette* in 2001, and in February 2008 there was the third annual Property Brands Conference. Recently, *Property Week* ran an online poll of the leading property brands. Most talk about names, logos and advertising but this is naïve as brand is a much broader concept, which should have a fundamental impact on every aspect of a business. Maybe this is because — as my friend and property law expert Geoffrey Lander pointed out — marketing people know all about brands and not much about property, and property people know all about property and little about brands.

There are numerous research studies showing that brand leaders dominate their market, command premium prices, enjoy superior profits and outperform their competitors financially. Brands can account for around 10–30% of the value of a company — which must be of interest when you consider the negative NPV of most property-listed companies.

Confusion about where brands fit in the property market

There is confusion between corporate brands being promoted to financial markets and investors and brands for particular developments, buildings, shopping centres or property-related services being promoted to occupiers and tenants. Agents are paid to "market" large developments — but do they really understand branding? Surprisingly, marketing — and branding — is not taught to surveyors.

Also, there is some confusion about the way to develop and promote B2B brands (where you target institutional investors or business tenants in the retail, industrial or office markets) and B2C brands (where you are target consumers — whether shoppers in retail centres or residents buying houses).

Building brands

Building a brand requires a management intensive, expensive and long-term strategy. Strong brands require a bedrock of rigorous analysis about a business' mission, market position and values. This requires carefully researched marketing strategies embracing market need and perceptions and integrated marketing communications campaigns, which ensure the differentiating message is delivered consistently through design, websites, PR, physical space and, most importantly, the behaviour of staff.

The increasing importance of the service element of property businesses means we have to help all staff act as appropriate "brand ambassadors". Staff must "live and breathe" the brand to deliver the "brand promise" on a daily basis. So we have to embrace training and internal communications in our brand-building strategies.

Emotional connections

Research shows that the market does not act rationally with regards to the value of property shares and there is a complex interplay between company-specific factors and average-sector discount but also some unidentified elements. It begs the question: To what extent does or could brand impact the irrational, sentiment issues?

Organisational buyers are no different to consumers who are influenced by both rational and emotional brand values. The brand insistence model defines five dimensions of brand — the emotional connection, the value delivered, relevant differentiation, awareness and accessibility. Brands represent trust and this is a precondition to loyalty. The view of academics Argenti and Druckenmuller is that reputation is managed reactively whereas brand is a future promise.

Leading property brands

A project manager at Segro (formerly Slough Estates) recently completed his MBA research dissertation on property brands by interviewing property industry CEOs. Five leading property "brands" emerged: Workspace, Land Securities, Brixton, Arlington and Helical Bar. In the recent *Property Week* brands poll, the winning brands were Land Securities, Londonnewcastle, Morley, British Land, CB Richard Ellis and Chainbow.

Land Securities used research to help pinpoint key issues, agree brand values and define key messages for all stakeholder groups. JP Morgan has described the contribution of its branding in reversing the trend against NAV to an earnings-based value. Stonemartin uses synergistic brand association with the Institute of Directors and its "IoDHub" flexible offices.

The strategic process at Brixton Estates drove its branding exercise, which has resulted in the development of a new service (B-Serve) for the customer service, leasing, asset management and estate/contract management elements of its business. The change from landlord mentality to a more customer-centric model of service provider required the recruitment of an almost entirely new team. The prize was an increase in customer satisfaction from 57% to 85% and renewals that have more than doubled since 2001.

Rok Property Solutions, a construction contractor, was formed in 2001 with its roots in nine offices, a few hundred staff in the West Country and a share price of around £1. Two years ago, Rok appointed a brand director to its board. As well as becoming one of a handful of property companies to gain a place in the *Sunday Times* Best Employer list, the company grew to 40 offices, over 2,500 personnel and a share price of around £5.

The bedrock of Rok's approach was customer feedback. The research is used to define service standards across the business. Garvis Snook, Rok's CEO, says that the core of the Rok brand is its people. "Good, commercial, motivated people. Selected for their quality skills and their "up for it" attitude to change and development". Every new Rok recruit — from the most junior technician to the most senior director — spends three days at a residential course absorbing these values and the Rok approach to customer service. Every employee is on the same benefits package. "We recognise that the Rok brand is created by every employee who comes into contact with customers — therefore, everyone must know what we are trying to do and is motivated to contribute to our success in every encounter."

When others in the property sector realise that brand means more than logos, we might see some significant development among property brands.

Building your own brand strategy

So what can you do to develop your brand? First, you must decide on just a few key values and messages that you want everyone — your partners, your staff, your clients and the market — to know and associate with your practice. This is much harder than you think it is. If you look at some of the truly great brands in the world they are so well defined it nearly hurts. For example, Disney focuses on "Fun family entertainment". Let us start by doing a health check on your current brand.

Table 4.5: The brand health check exercise

The healthy brand criteria	Healthy	Could be improved	Unhealthy
Based on a proposition of genuine substance and value to the target client?			
Communicates a clear and powerful brand definition?			
Communicates a clear "emotional charge"?			
Communicates an attractive and relevant personality?			
Wins, builds and retains client loyalty?			
Is well known by the target client?			
Is held in high esteem by the target client?			
Communicates a unique match between the firm's capabilities and the client's needs?			
A source of competitive advantage?			
An investment of increasing value that others will want to own?			
Maintains relevance over time by evolving response to changing client expectations and perceptions?			
Increases the profitability of the business?			
Consistent with business strategy?			
Makes sense within the business' brand architecture?			
Provides a protective halo for growth strategies?			
Uniquely positioned in the market and creates a relevant space in the client's mind?			
Communicates and demonstrates a clear sense of value?			
Interacts consistently with the client on as many fronts and on as many occasions as possible?			
Cements the brand definition into the client's mind through interactions and positive associations?			
Managed and supported consistently over time?			
Have values that can be applied consistently to all parts of the marketing mix and through all media?			
Makes people want to get their hands on it?			

Source: Extract from "Understanding brands" *The Sunday Times Creating Success* book by Peter Cheverton

I suspect that you realise that you need to do some work on your brand (or you have decided to opt for a rather different marketing strategy). It might be worth doing an exercise with some of your partners where you consider the various aspects of your brand and work up to a possible branding strategy. The exercise in table 4.6 might help.

Table 4.6: Create your brand exercise

1. Complete for your firm

Features	Benefits	Values	Personality
Who is your firm? What does your firm do? What does your firm offer?	How are these things used and/or valued by your clients?	What are your partner and staff's strongly held values and/or beliefs?	Adjectives to describe your firm's "personality"?
Brand promise Differentiated benefits that are relevant and compelling to the "buyer" (functional, experiential, emotional or self-expressive)			

2. Compare with similar firms/competitors

Points of similarity with others	Points of difference with others	
Relative positioning against others:		
Brand image		
Firm's self image	Projected image to others	Received image — what the market perceives
Brand essence How best to summarise everything into a single compelling statement (3–5 words) eg, Nike — Genuine Athletic Performance		

I have to admit that branding is more difficult than you think it is. Even as a marketing expert with nearly 30 years' experience, I still do not feel that I am an expert at branding. As with most professions, the marketing profession has a number of different branches and specialisms — and branding is one. If you have plenty of money, you could hire a branding expert. Or you could invest in some of the entry-level branding books I have mentioned in this chapter and have a go yourself. But remember, as with most marketing, often the journey of discovery is more important and more valuable than the final destination. I am still very proud of the "Real People. Real Solutions. Real Estate" strapline that summarised the brand essence on Weatherall Green & Smith nearly 10 years ago. And it took around two years' analysis and immersion in the firm's information and its clients before inspiration struck me one evening in the bath.

Checklist and actions

- Look at the methodologies for your present work to see if you can package anything that is easier to promote to potential clients.

- Consider how you need to innovate in order to develop new products or services for the future.

- Think carefully about a suitable pricing structure that fits your overall strategy, preserves your existing client relationships and positions you appropriately in the marketplace.

- If you think that brand will be important to you, complete the various exercises — along with your partners — so that you agree about the main aspects of your firm's brand.

Decisions, decisions — what makes a good marketing strategy?

As I mentioned above, one of the trickiest parts of a good marketing strategy is persuading all of your partners to agree on which markets and services to focus on. Too many firms spread themselves too thinly — they try to develop a strategy that gives equal emphasis to the interests and preferences of all the partners. This may keep the partnership harmonious but it does not produce the sort of focus you need for a good strategy. Furthermore, you risk confusing the market by sending out too many different messages and failing to emphasise the two or three key points that you are most keen (either through a market or service focus or through a branding strategy) to promote — to maximise your competitive differentiation.

It may help you to think of appropriate strategies by considering the following. Pull strategies are aimed at end users (consumers, the residential market) to cause them to demand the product from suppliers (or seek you out on the High Street or internet). Branding is the most effective method here. Push strategies are aimed at sales channels (eg, intermediaries), supporting and encouraging their efforts to increase sales. Profile strategies address the way an organisation is perceived by its various stakeholders.

This is difficult because you know a good marketing strategy when you see one but cannot describe it in conceptual terms. However, the academics would suggest you consider the following.

Nine elements of a good strategic marketing plan:
- Clarity of scope.
- Definition of intended competitive advantage.
- Internal consistency and synergy.
- Degree of uniqueness.
- Congruence with the external environment.
- Consistency with the organisation's objectives.
- Acceptability of risk level.
- Feasibility within the organisation's resources.
- Provision of a level of guidance to tactical activity.

Now that you have done all that hard work, I think you should to get it summarised into a simple marketing plan so that you remember what you did, what you discovered and why you made the choices that you made. This will make it easier to communicate what you are doing and what you want others in the firm to do.

To plan or not to plan? Simple marketing plans

For those of you who are already plan-weary, I also offer a simpler version:

SOSTAC
- Situation
- Objectives
- Strategic
- Tactics
- Action
- Control

Or even — 3M — (wo)men, money, minutes.

Campaigns

Within your strategic marketing plan it should be clear what markets are being targeted (and a really good plan will indicate what particular organisations you are targeting in that market if you are in the commercial space) and with what services. You will also show how, over time, you plan to use the various marketing communications or promotional activities to raise awareness, develop understanding, create or reinforce conviction and then encourage action (ie, get in contact with your firm in some way). These activities, if properly selected and integrated, form a campaign that might last anything from three months to a year or more.

This way you combine the impact of all the different activities and can measure their effectiveness in achieving your aims. Otherwise, you are faced with the difficult task of trying to measure the effectiveness of individual activities — such as press releases, seminars and hospitality — which is notoriously difficult — particularly as most clients make their purchase decisions based on a number of different marketing activities and some very good sales activity. You should consider what you are trying to achieve in terms of campaigns — where you blend the various tactical activities — as they are described in chapter 5 — into a coherent and integrated campaign. This way, all the different tools will work together to reinforce each other and act at the appropriate points in the selling cycle.

Table 4.7: Illustrative campaign plan

Programme/Activity	Mar-09	Apr-09	May-09	Jun-09	Jul-09
Awareness					
PR research		▓			
Press launch and interviews			▓		
Ongoing articles					
Internal communications					
Project team briefing	▓			▓	
Lunchtime staff workshops		▓			
Departmental meeting updates				▓	
Lead generation					
Database compilation/research		▓			
Newsletter production			▓		
E-mail campaign				▓	
Seminar programme					
E-mail campaign			▓		
Seminar programme				▓	▓
Sales programme					
Telemarketing follow up					▓
Appointments					
Presentations/pitches					

Source: Kim Tasso

How much? Budget questions

The marketing plan has been produced and we all sigh with relief. But then you have to decide how much money to allocate to your marketing budget. How long is a piece of string? Or, more likely, how much was last year's piece of string?

It is true that many firms simply take whatever they spent on marketing the previous year and add (or remove if times are hard) a percentage in order to create the marketing budget. Others might be a bit more sophisticated and look at the relative spend on things such as entertaining, exhibitions, advertising or publications. But this really misses the point.

The right way to develop a marketing budget is to look at what you are trying to achieve and to develop a budget from a zero base. And check that you are likely to receive the appropriate return on that investment. We call this the "objective and task" method of setting a marketing budget. The 10 steps are as follows:

1. Define the marketing and promotions objectives.
2. Determine the tasks to be undertaken.

3. Build up expenditure by costing the tasks.
4. Compare results against industry averages.
5. Compare results as a percentage against sales.
6. Reconcile differences between steps 3, 4 and 5.
7. Modify estimates to meet company policies.
8. Specify when expenditures are to be made.
9. Maintain an element of flexibility.
10. Maintain actual results against those forecast.

Since people always ask me, I suggest a rule of thumb along the lines of 2–4% of your gross fee income for your marketing budget. Obviously, this amount will be more if you have done little marketing to begin with or have particularly aggressive goals. And less if you cannot afford very much.

Make or buy? Acquiring marketing expertise

If you do not have any marketing expertise (or time) then you may need to think about getting some. Many property firms think that they can hire a relatively junior marketing person, and then return to doing what they do best — serving their clients. But this is wrong.

First, a junior marketing person will be unable to work effectively without a good strategic plan and a deep understanding of your business (how much time and effort do you invest in helping your marketing person learn about your markets and services? When I first worked for a law firm, I had to share an office for six months with one of its most senior corporate lawyers. I learned a lot about what lawyers do. Then I repeated this exercise when I first started working within a property company (although it was a corporate real estate consultant I shared an office with there).)

Even if they have a good understanding of your firm, its markets and its services and you have produced a brilliant plan that is not the answer. Because the in-house marketer will need lots of time with the senior management and the surveyors and agents throughout the business in order to do their job effectively. Therefore, having an in-house person increases the amount of time everyone spends on marketing and business development rather than reducing it.

Also, very often the in-house marketing person is little more than an administrative assistant. Typing information into databases, chasing surveyors for articles to go into newsletters or property magazines or organising parties. This is all good stuff but it is administration and not marketing.

An alternative solution might be to hire an external consultant. And these, as with surveyors and agents, come in all shapes and sizes with different strengths and specialisms. So if you hire a strategic consultant that's what they will do, if you hire a public relations expert then that's what they will do. So choose carefully.

I should mention here that *Media Relations in Property*, the other book I wrote for EG Books, devotes considerable space to explaining how to develop a brief, how to review and assess tenders from consultants and how to manage them (and measure) them effectively on an ongoing basis. The same process can be used for hiring other types of marketing specialist.

Keeping your eye on the ball — monitoring

You have the plan. You have the expert resources. You have all the partners and other fee-earners on board and doing their bit. But is it working?

A good marketing strategy should guide your day-to-day decisions and should help you keep track of what is happening and why and whether it is producing the anticipated results. Therefore, use the plan to undertake a regular review of progress — both in the process of marketing (what is happening) and the results (what occurs as a result).

If you find that things are not going according to plan, then you need to revisit your analyses, assumptions, aims and strategies to learn why not. And to make changes if required to ensure that things start to move in the right direction again.

Allocate sufficient time at each monthly or quarterly management or partnership meeting to review the milestones or critical success factors in your strategic marketing plan and have the relevant data to hand so that you can check that you are on course. Otherwise, allocate time to consider the alternative approaches that should be adopted.

Checklist and actions

- Assess whether your new marketing strategy is good enough using the criteria in the chapter.

- Produce a marketing plan — even if it is a very short and simple one.

- Discuss the proposed plan with your partners and refine it as necessary.

- Communicate the main elements of the plan to your staff — and obtain their input and ideas and involvement.

- Organise the main activities into coherent campaigns for both internal and external communications.

- Set and approve an appropriate marketing budget.

- Acquire the relevant marketing expertise — through training yourself and/or your partners, hiring an in-house marketer or commissioning a marketing consultant.

- Put in place the relevant systems and procedures so that you can measure both progress and results and make adjustments when they are required.

- Pat yourself on the back for getting through the tricky process of strategic marketing planning!

Case study: Cluttons — the strategic repositioning of a brand

Cluttons is an old and well-established property services firm. Founded in 1765 by John Clutton, there have been family members in senior positions for seven generations. Until 2003, the firm was headquartered in a striking Grade I listed building in Berkeley Square with two mews annexes. It had a good number of establishment clients, whom it had served for a long time; in some cases, for many decades.

And therein lay the problem. Although respected, the business was seen as a bit old fashioned. The headquarters had grown with the business and its complex physical layout supported the more traditional approaches of a management and general practice property partnership. People had their own offices and were generalists in their fields with their own client relationships.

Externally, things were also making it difficult to operate to the standards on which the firm prided itself. Customer service expectations were growing; it was becoming increasingly hard to keep up with regulatory changes, and competition was encroaching on its traditional areas of business.

As a result, and as is often the case, a sense of dissatisfaction gradually built within the younger members of the firm. The opportunity to do something about it came following the 1997 merger with Daniel Smith and later the natural retirement of several senior people. New blood decided to take the bull by the horns and modernise. A critical element of this was to bring strategic marketing into the management of the business.

The Cluttons brand

As with most professional services organisations, Cluttons relies on three key elements to sustain its business:

* Knowledge and skills, which do not merit an in-house provision by clients.
* Its brand — not in the Coca Cola sense but in terms of its reputation.
* Relationships with clients that enable them to be a preferred supplier.

Of these, the brand was of particular strategic importance to Cluttons and the strongest basis for long-term competitive advantage. In the absence of a tangible product, the brand provides an assurance of quality. In addition, for the more transactional agency services, the brand substitutes for longer-term personal relationships.

At the time of the decision to modernise, however, the brand reflected the company as it had evolved: traditional, sound professionalism, with a "country-house" feel. The issue was how to modernise it while retaining the core values of experience, solidity and integrity.

Cluttons recognised that repositioning a brand in commercial markets was not a simple matter of logo redesign and lots of advertising. Indeed, the logo redesign is often more important as a way of symbolising change to the employees than it is for the market. More important is the consistent delivery of the brand's values. As an early marketing director of Orange once said: "To be Orange on the outside, we also have to be Orange on the inside."

In other words, modernising the brand had to be underpinned by a modernisation of the business. In this, the senior partners introduced a number of changes. On their own, none of these would create a change in brand perception. Collectively they led to a significant repositioning.

Creating the foundations

Most obvious in the brand modernisation process was the move of the main offices from Berkeley Square to Portman Street. Apart from being a more contemporary building, it enabled things to be done differently. At Berkeley Square it was said (apocryphally) that visitors sometimes went home because the person they had come to see took so long to find their way to reception. At

Portman Street the more open layout meant that visitors could be warmly welcomed, see what was going on, feel more in touch with the business, and be met much more quickly.

The deliberately open-plan design also served to emphasise the move from individual relationships to service based on complimentary teams of specialists. Senior partners were seated among their staff to improve visibility and dialogue and to enable them to role model different ways of working.

A marketing department was created to bring specialists into the presentation of the business and to support customer-focused activities, such as tendering and relationship management. At the same time, new internal communication practices were introduced. There were regular presentations to the staff, team meetings, consultation exercises and much greater levels of transparency. These initiatives supported enhanced professionalism and meant that staff were much better informed about the firm, its ways of working and its services.

With the addition of a new IT system that enabled an intranet, remote access, homeworking and hot-desking; the look and feel of the business evolved significantly.

External perceptions

From the point of view of the market, the business began to lose its traditional general-practice "country house" image. Thus, presentation material became more uniform as it conformed to centrally determined standards and formats that reinforced the new image. Improved transparency and communications enabled staff to be seen as more in touch with what was going on and to talk more knowledgeably to clients about the range of initiatives that were being undertaken.

As an example, the new IT infrastructure enabled innovative services such as secure remote access from any location over the internet. Cluttons staff, business partners and, of course, clients can now login and share relevant business information. Further application developments will provide clients with greater visibility of their projects.

The quality aspects of the brand were retained with the help of the mantra "the right person with the right expertise", which ensured that clients received the best level of service and experience. They supported this with a focus on those areas where they had an advantage as a result of expertise, experience, established relationships or presence.

One tough decision that was made in 2008 to ensure that activities were aligned with the modernisation of the brand was the disposal of the Rural Division. Although profitable, it did not align with the strategic emphasis on commercial sectors. In addition, it did not help the business move away from the "country house" image it was trying to abandon.

The rise of the telecoms team

As the business became more market-oriented and the modernised brand image consolidated, a whole new raft of opportunities became open to it. The rise of the telecoms team within the Commercial Division is one example of this. To give some idea of scale, over a 10-year period to 2008, revenues rose from about £15,000 to £2m per annum.

Stemming from a long-term relationship with the BBC, the company had developed expertise in many areas that were of concern to the emerging and fast-growing wireless telecoms market. However, expertise on its own was not enough to achieve the dominant position in the market that Cluttons now holds.

To begin with, modern fast-growing companies often look for equivalence in their suppliers — they prefer to work with organisations they feel comfortable with. The modern more dynamic brand that Cluttons had developed fitted this bill for the greater part of the market.

The new team approach incorporating a range of specialists meant that the business could meet almost all the property-related needs of the telecoms companies. In addition, where necessary, experience was recruited from relevant parts of the telecoms industry to supplement home-grown expertise, ensuring that Cluttons' standards were maintained.

Final word
Cluttons has embraced strategic marketing as a key organisational discipline. It recognises that marketing in the broader sense is not simply something that marketing departments do, but is a business-wide activity led from the top. The trick has been to build on the firm's marketing assets; in this case, the strong reputation or brand that the firm's 21st century leaders inherited.

Key to this has been a different structure and a way of working that matches the markets the firm seeks to serve. The team approach facilitates higher levels of individual specialisation and innovation, and enables the firm to match the ebb and flow of what is now a dynamic property market.

For the future, the firm has a strategic plan with goals stretching to 2020. Again, the approach will be to build on their assets and to protect their brand values of experience, solid professionalism and integrity.

To borrow from the Orange marketing director quoted above; to be the new Cluttons on the outside they have to be the new Cluttons on the inside as well.

Case study: Pellings — strategic market planning

Pellings LLP is a multidisciplinary property and construction consultancy. Founded in 1977 the firm has offices in North and South London and approaching 100 staff including 15 partners, seven of whom are senior members of the LLP. It operates mainly in the public sector, particularly the public housing and education sectors, throughout Southern England.

A period of change
In 2003, the firm was at the stage where its founding partner was retiring and it was introducing new partners. The practice had grown from its traditional

building surveying roots and become a multidisciplinary practice. However, there was a lack of strategic thinking about the practice and a latent potential among the partners for cross-selling.

At the same time, the practice was advised that it was underperforming financially compared with similarly-sized direct competitors. This gave the partners the impetus to undertake a strategic market planning exercise, which helped transform the business over the next four years.

Business planning

For the first time, the management team decided to set explicit targets for the firm that reflected their desire for a competitive and profit-orientated business. They also engaged the support of a strategic marketing consultant, who helped them to develop their marketing plan.

Using a matrix of sectors and services Pellings undertook a market audit, which made them review who their existing clients were; where their existing business came from and what the profile of clients were likely to be in the future. The process was about more than simply finding out who their clients were; and a series of customer surveys and questionnaires helped the management team understand why clients gave them work.

According to Richard Claxton, senior manager at Pellings, this was one of the factors that really helped make the plan relevant:

"The process gave us the opportunity to ask clients about our services: what they liked about us and what needed improvement, and the services they would require in the future. It's vital that professional services firms do this as often as they can and the business plan put the responses into context."

The outcome of the market planning process was that Pellings identified and set challenging — yet achievable — growth targets for their existing major and "growth" sectors (ie, sectors where they had little or no presence) and similarly set targets for their major and "growth" services within each sector.

Success achieved

The underlying target was to bring Pellings into line with its competitors in terms of profit and turnover per partner, and this has been achieved over the four years of the plan. The firm has grown its turnover by nearly 40% over the period of the plan. Profitability has also improved and is now in the upper range within its peer group.

The business has also launched a new project management service within the period of the plan in response to the customer demand for this service identified through the process.

A sector-based approach to marketing and business development is seen as the main driver of this achievement. The firm operates five sector groups and a cross-sector sustainability focus group. Each group has an annual marketing plan, and meets regularly. However, according to Richard Claxton, more importantly, the sector groups have broken down the barriers to joined up thinking across disciplines, particularly at a junior partner and associate level:

> "The sector groups have brought a number of staff forward and allowed them to develop their skills. It has given them a broader understanding of what the business can offer clients overall and I see joined up thinking going on every day. The sector groups also give our staff confidence that we know where we are going and are investing in the future of the business."

Improvements and challenges

The strategic market planning process has also acted as a trigger for improved internal communications through an annual staff conference where the plan is reviewed, and attendees are asked to become involved in some aspects of the plan.

Richard is clear that engaging the staff in the plan is key to its success:

> "We want our people to operate as one business and 'Proud to be Pellings'. Undoubtedly, there is an element of scepticism that we needed to overcome. Our staff needed to learn to trust one another and deliver for one another, and this has been a challenge."

The targets identified in the business plan are also incorporated into the firm's quality management system and performance is regularly measured and reviewed — although there are no penalties in place for sector groups or services that fail to achieve their targets. Richard says:

> "We recognise that the various parts of the business operate in different cycles and the plan is not set in stone. Measurement is about continuous improvement and helps direct our efforts to sectors or services that are underperforming or equally where we may be over committed. It is essential to continually review the plan's performance throughout its life."

Lessons learned

Lessons have been learned by the firm along the way and not everything has been plain sailing.

In the first instance, they ran the business plan in parallel with their financial year, but discovered that they did not have access to the required financial information at the right time. They now run their business plan by calendar year and this gives the management team access to the information they need to review their performance.

They also considered setting targets based on the shareholding of partners and letting the targets in the plan drive staffing levels. However, both ideas were rejected:

> "We quickly realised that if we allocated targets to specific equity partners we may fail to engender the ethos of co-operation we were aiming to foster throughout the

business. We also realised that it was impractical to make significant changes to our resource levels simply on the basis of the business plan and our plans now take more cognisance of current staff levels and the availability of well trained resource in the open market. This makes it more difficult for our 'growth' sectors and services that have less resources available to them, but you need to balance your growth plans with the risk you are able to accept."

Richard and the management team at Pellings now consider a detailed business plan essential. They are about to publish a plan for the next three years and are looking forward to using it to help them stay focused, communicate and deliver continuous improvements for their clients and staff.

Tactical marketing (promotional strategies)

Roadmap

- Throw your money away!
- Apples and pears — marketing properties is not the same as marketing professional services
- Logos, straplines and visual identity
- Rules! Rules! Rules! A consistent housestyle
- All mixed up — marketing and communication mixes
- A big DRIP
- Aunty AIDA
- The digital age is upon us
- Database nirvana
- Awareness versus relationship marketing
- Choose your weapons
- A–Z of promotional or marketing communication tools
- Better briefs — choosing and using external specialists
- But what *really* works? Measuring effectiveness

Throw your money away!

So now that you have done all the brain-hurting analysis, planning and strategic stuff it is time for the fun part. If you have not done the tough planning stuff, then the money you spend on tactical marketing will be wasted.

Having been a judge on The Company of Chartered Surveyors annual PAMADA (Property Advertising Marketing And Design Awards) before they became the Property Marketing Awards, I have seen a wide variety of tactical marketing campaigns from all areas of the property sector. Generally, in the early days, I was pretty disappointed and saddened. Imagine, a huge conference table covered with over 30 expensive brochures and finding it difficult to pick one out that is not the same as all the rest. Or trying to find just one that shows a glimmer of inspiration and interest. Imagine reviewing what is supposed to be the "cream of the crop" and wondering how on earth the firms managed to think that what they were submitting was in some way special. But then again, the awards were established by the Company of Chartered Surveyors to raise standards and, over the years, they really did. Yet even in the later years, as the design improved significantly, the submissions rarely showed SMART objectives or reported on the effectiveness and results. Hopefully this chapter might provide some more interesting and exciting submissions for future judges as well as help firms to produce more effective marketing in the future.

Many years ago, when marketing was not very developed in the professions, one rather gruff partner referred to the marketing department as "the colouring in department". The property industry (shame on you) was also famous for referring to the marketing team as "the pretty girls who organise all the parties". Also, I remember attending a pitch for the PR account of a major property firm where the choice of the winning agency was based on the attractiveness of the young (female) account executives.

Apples and pears — marketing properties is not the same as marketing professional services

For many years, those in the property industry were pretty damning about the marketing department. They argued that clients paid them very significant sums of money to undertake the marketing campaigns for their developments. And clients did. The agents spent loads of money on a fairly standard "mix" — a glossy brochure and a humourless advert in *Estates Gazette* or *Property Week* that they could then hopefully submit for an award. Then they sold or let the property with a couple of telephone calls to their mates in other agencies. Job done. But this is not marketing, it is money wasting at worst and ego-brochures at best. As you can tell, I used to take a bit of a dim view of the majority of property marketing but things are improving.

I am happy to receive examples of great marketing. In fact, I would be delighted to have to write a retraction in *Estates Gazette* at some point in the near future explaining about how excited I was to discover some first-class marketing in the property industry at long last.

The point I am trying to make here is that the skills and tools to market a property are quite different to the skills, tools and approach to marketing professional services. I even produced a table to show you the differences in the previous chapter on strategic marketing.

One of the reasons that this book was written was because, although a large part of many agents' work is marketing and selling properties (not marketing their services, which they are often resentful of devoting time to), a formal marketing education is still not part of surveyors' APC training.

Logos, straplines and visual identity

Logos, straplines and visual identity are part of your brand strategy (see chapter 4 on strategic marketing for more information about brands) but not the be all and end all of it. Most people get them confused or do not realise the importance of the various elements or of having some standards (as encapsulated in the visual identity) that ensure all materials — whether printed or electronic — have the same look and feel and clearly come from the same "stable".

I wrote about this extensively in chapter 4 but it is probably easiest to explain with an example. One of my favourites was when I was at Weatherall, Green & Smith in 2001. We "refreshed" the design so that it moved from this:

Figure 5.1: Old Weatherall Green & Smith logo

Weatherall
Green & Smith

To this:

Figure 5.2: New Weatherall Green & Smith logo

With the W in the logo there are two colours — the traditional and existing blue but an additional green to give us more flexibility and also to link into the greener aspects of the property business. We shrank the "Green & Smith" further — knowing that it was only a matter of time before we stopped using it altogether and recognising that most people called us "Weatheralls" anyway. Quality was an important aspect of the business and the green tick is a reminder of that. The overall shape of the logo (a rugby ball) was seen to represent the more human and friendly side of the otherwise quite "cool" and rather straight words. Our brand strategy focused on the quality, approachability and problem-solving approach of our people.

I can also take credit for developing the strapline — "Real People. Real Solutions. Real Estate". We were trying to link the firm's extraordinary experts and their commercially minded business solutions and make it the central platform of the firm's differentiation.

It must have worked as the ideas were adopted when the firm was acquired first by Auguste Thouard and then Banque Paribas. And "Real" became incorporated into the new name — First as Atis Weatherall and then Atisreal — the strapline "Real people. Real market coverage. Real solutions" has evolved to show the European-wide group's strength across Europe.

Figure 5.3: Latest logo

Rules! Rules! Rules! A consistent housestyle

But visual identity goes beyond these elements as part of your branding strategy. You need to think about the "look and feel" of your other materials. Ideally, these should be documented in a corporate-style manual and someone, somewhere should be "policing" that all materials produced comply with the standards (even in the Brighton office, who swear that their local reputation and market is different!). Are your brochures of a standard size? Do you use the same pantone colours in print as you do on screen? Are the fonts you use in your presentations and documents the same as those on your website? Does the photographic style you use conform to certain conventions — or do you mix up different types of photos?

Photos are incredibly important. Most property businesses are only differentiated by their people. But when your brand says you are, for example "Accessible. Friendly. Pragmatic. Young" do your photos say "Serious. Distant. Grey. Old"?

For your corporate identity you need to check that you have clear rules of engagement for the logo, the strapline, the colours, the typography, the tone of voice of the copy, the literature size and formats, the photos and any illustrations or graphics.

Colour is an interesting issue. Many property firms stick with a nice safe but dull dark blue. Some will go with a sort of Harrods green — traditional, safe, uninspired, conventional. But there are so many colours and they can signal some interesting messages:

- Non-primary colours are more calming than primary colours.
- Blue is the most calming of the primary colours, followed closely by a lighter red.
- Yellow invokes cheerfulness (houses with yellow trim or flowers sell faster).
- Red trim is used in bars and casinos because it can cause people to lose track of time.
- Red makes food more appealing and influences people to eat more.
- Forest green and burgundy appeal to the wealthiest 3%.
- Orange is often used to make an expensive item seem less expensive.
- White is typically associated with cool, clean and fresh.
- Black is associated with elegance and sophistication.

Just today (7 January 2009) there was a research study by Dr David Lewis at the University of Sussex reported in the media that showed those exposed to a colour completed tests 25% quicker — even strength was greater. The tests revealed blue promoted calm. Purple was also relaxing, but only for women. Blue and green made the men feel happier while blue, purple and orange raised female spirits. Red was the colour least likely to make people feel happy.

Checklist and actions

- Did you do a business plan and a strategic marketing plan, as advised in the previous chapters?

- Have you put all of your own and your competitors brochures onto a conference table and taken a good, hard look — from the clients' point of view? What did you learn?

- Have you thought about the differences between how you market your clients' properties and your own firm and services?

- Have you agreed on your brand strategy and aligned it with your corporate identity?

- Do you have a proper identity/style guide showing how you use your firm's name, logo, strapline, typography, photos and copy?

- Who polices the use of your identity so that it is always consistent?

All mixed up — marketing and communication mixes

Marketing people talk about mixes. They start with the marketing mix — this is the 4 (or 7) Ps addressed in chapter 4 on strategic marketing. Analyse the market Place, refine the Product, set the right Price and then develop an appropriate Promotional mix. This promotional mix, sometimes called marketing communications, is the focus of this chapter. It is the bit that you must not do until you have sorted out all the strategy and planning stuff mentioned previously.

The marketing communications mix comprises five main categories of communication and each has its strengths and weaknesses (Table 5.1).

From table 5.1, we can see that while advertising might be very good for communicating about a specific property, it is rather inadequate when trying to convey the detailed information about a professional property service to a targeted audience.

A good tactical marketing approach will select and blend the different tools to achieve the specific aims set out in the strategic marketing plan promoting the right mix of product at the right price to the right (market)place with the right promotion.

Table 5.1: Comparing promotional tools

	Advertising	Sales promotion	Public relations	Personal selling	Direct marketing
Ability to deliver a personal message	Low	Low	Low	High	High
Ability to reach a large audience	High	Medium	Medium	Low	Medium
Level of interaction	Low	Low	Low	High	High
Credibility given by target audience	Low	Medium	High	Medium	Medium
Cost	High	Medium	Low	High	Low

Source: Chris Fill

A big DRIP

One of the most important models I use when teaching young marketers about effective marketing communications is DRIP. This stands for:

- Differentiate — Does your communication (both the message and medium) differentiate your firm, product or service from competitive offers?

- Reinforce — Does your communication reinforce key strategic messages or your overall positioning? Does each communication reinforce the effect of previous and parallel communications?

- Inform — Does it provide the right amount, level and type of information that is needed by the recipients?

- Persuade — Does it cause them to take some action — for example, visit a website, call to make an enquiry, send off to obtain some information or arrange a meeting?

It is good to ask yourself what you are trying to achieve with each piece of marketing activity. And do not forget to integrate your marketing communications strategies and campaigns with your sales strategies otherwise you could use the wrong tool at each part of the sales cycle.

Aunty AIDA

Another old but incredibly useful tool that I use is AIDA. No not the famous opera. AIDA helps you guard against marketing activities that are too inwardly focused. If you use the word "We" then you are simply not going to get them interested. They are interested in stuff about "You" — the reader. You need to focus on what is of interest to the target audience. The benefits and value to them. This means getting into their heads and seeing things from their perspective. Chapter 6 on selling devotes considerable time to the concept of empathy, as it is fundamental to all aspects of marketing, selling and relationship development.

I use AIDA to test out marketing materials — whether they are direct mail letters, brochures or even e-alerts — to see if they conform. It stands for:

- Attention — Does the visual appearance of your item grab attention? Does the copy immediately engage with the audience on their agenda? Does it make them sit up and pay attention?

- Interest — You have them hooked, so how are you going to keep their attention? They are receiving around 8,000 advertising messages each day so you are going to have to say something meaningful if you want them to devote the few seconds it takes to get them reading the first paragraph.

- Desire — Have you now generated enough desire to get them to start thinking about what they really want to do about a particular issue or problem? If there's no motivation, then there will be no action.

- Action — You got their attention and developed their interest and created a desire for some change. Now you need to give them a nice, low commitment action to undertake NOW. A visit to the website for more information, an email to obtain an information pack or more details of a seminar.

A similar acronym to help you structure your communications — which is particularly useful in sales documents is — Problem, Promise, Proof and Price.

The digital age is upon us

Previously (ie, about five years ago), the majority of promotional work used tangible, hard copy tools. Brochures that you could hold, samples you could sneak into your pocket, letters and invitations that came through your letter box. While these tools made an impact (assuming they reached the hands of the people for whom they were intended) they were time-consuming and expensive to produce and took some time for them to reach the right hands.

Then the internet exploded onto the scene and reinvented marketing. All the old rules became redundant and all the marketing tried to make sense of this brave new world. I have to say that the property industry was — and to some extent remains — rather resistant to all things digital. Technology has rewritten the rules and changed the face of marketing fundamentally. And it is still changing. So apologies if things change or move on further between the time of writing this and when the book becomes published.

There are various sources to tell you about the impact of the internet on marketing and advertising. For example, the Internet Advertising Bureau reported that in 2003 online advertising was the smallest sector yet, in 2007, it grew nine times faster than entire advertising market (£17.5bn) to become the third-largest at £2.8bn — and is expected to overtake television advertising in 2009. Nearly £600m is now spent on online display and embedded format advertisements (banners, skyscrapers, video). Over 35m people in the UK have internet access and the average broadband user spends 16 hours a week online. They also report that social networks are driving audiences, but spend on these sites is relatively low. I recently wrote a piece of research into the grey market and was surprised to discover that the over 50s are bigger users of the internet than the younger generation.

Checklist and actions

- Do you consider the strengths and weaknesses of each marketing communications tool before deciding to use it for a particular campaign?

- Do you think about how each marketing activity differentiates, informs, reinforces or persuades?

- Have you tested your recent marketing materials with the AIDA tool?

- Have you counted how many times your marketing materials (including your website) refer to "we"?

- Which of your partners are digitally savvy?

- How can you increase the level of internet awareness within your firm?

Database nirvana

As a result of the cheapness, immediacy and speed of email communications, the need to have an accurate database up and running (and constantly updated either centrally or by all the people in your firm) becomes even more important.

Yes, I know that you have been trying (unsuccessfully) for many decades to get it sorted out. But let me tell you a secret. It is not about the software system. It is not about the IT department. It is not even about the marketing department. The problem with databases in property firms is the partners and fee-earners. They do not want to share their information. And they simply cannot be bothered to invest the necessary time and effort in entering their client and contact information and keeping it up to date.

The majority of marketing activity relies on accurate and comprehensive data. So if you don't have a database STOP. Do not pass this point. Go away and sort it out. Start simple if you like, by selecting the firm's 100 most important clients, or ask each partner to get the information right for just their top 50 clients and contacts. I don't care what you do or how you do it — but please DO SOMETHING!

Also let us think about what sort of data you need in that database. Obviously a name, position, organisation, address, telephone number and email address is a good starting point. But you also need data on a number of other issues if you are to first do a segmentation analysis (see chapter 4 on strategic marketing) and second if you wish to target your communications accurately to match the needs and interests.

You might consider using some codes for each contact or organisation on your database so that you can easily and quickly extract the relevant list without the usual 66 sheets of printout that the partners need to mark up for the next newsletter mailing or invite to a reception. The sorts of codes (and you need to balance ease of use with targeting accuracy) might include:

- **Industry sector**
  Is the commercial organisation concerned in the property industry, the educational sector, agriculture and so on? Some of the most successful marketing focuses on showing your experience and understanding of a particular sector's needs. Show that you can speak their language and that you have property solutions that are tailored to the particular requirements and regulations of their sector.

- **Size**
  Large, medium or small-sized organisations are likely to have very different property needs. Thus you need to tailor information and events to their specific interests. You might have a similar code to indicate those that are changing fast or those that are relatively stable.

- **Location**
  You will have the postcode and some of you have the software that enables you to extract relevant towns, districts or regions. But if you do not, then it might be worth using codes to indicate the particular office or region where you have a particular programme (eg, if you go to county shows you might want to alert all those in that county that you will have a stand there).

- **Property type**
  Some organisations will just be interested in offices whereas others will be interested in industrial and residential developments as well. Maybe you can do this by linking the partner who "owns" the contact with the relevant department?

- **Type of client**
  Maybe you have long-established key clients who are part of a special relationship programme? Perhaps you have a list of hot prospects or targets that you want to ensure receive regular information? Maybe there are clients where you know there is a potential for cross-selling? Alternatively, you may have some clients who subscribe to just one of your services and have indicated that they do not want other information?

- **Contact log**
  If your database is more than a contact manager and has facilities for CRM (Client Relationship Management), then you may be able to keep a whole host of information such as when you last contacted them, the time you are due to contact them again, their particular interests, background information about their organisation or the relationship, the various people at your firm who they know, the other advisers that they use or even information about their property portfolio. Sometimes, firms have systems that allow them to integrate their contact information with Microsoft Office software, so that your contact list automatically updates your daily schedule with reminders and such like.

- **Areas of interest**
  You may have a list of topics (eg, on your website) where clients, contacts and referrers can indicate their areas of interest — for example, human resource issues, health and safety, corporate real estate, relocation, investment. Perhaps you have a list of newsletters to which they can subscribe. It is much better to have these things coded in one central system then relying on a medley of spreadsheets and Word documents strewn around the firm. This means there is much less chance of duplicating effort or, worse still, omitting a key client from an important alert.

For other ideas — particularly if you are focusing on the consumer or residential market — you should look at the section on segmentation in chapter 4.

You should also be familiar with some of the legislation surrounding databases. The most important of these is the Data Protection Act. This means that you must take good care to ensure that the information is up to date and only used as the individual has agreed is appropriate — no sharing without permission. You should also know that people are entitled to see what information you hold about them, so take care with any personal comments.

Another critical piece of legislation relates to email marketing. You are required to ensure that people opt-in to receiving emails. There are hefty fines for those who email indiscriminately and it also does significant damage to your reputation. So make sure you can keep a note of the date on which people opt-in to receiving emails from your firm.

Checklist and actions

- How good is your client and contact database?

- How able are the partners and staff (and the procedures and training to support them) at entering and maintaining their client and contact information?

- Do you have the right codes and tags to enable you to extract the correct list at the right time with ease?

- Does your database have the potential to become a more sophisticated CRM system?

- Who has ultimate responsibility for maintaining the database? And the data quality?

- What sort of project (aims, resources, time scale) would you need to sort out your database?

- Are you familiar with the Data Protection Act and email marketing regulations?

Awareness versus relationship marketing

Despite all the different tools in the marketing communications toolbox most people use them to achieve one of two jobs — either they are there to help raise awareness of your firm, its services or its people or it is designed to initiate and develop a relationship — whether this is with prospects or existing clients and referrers.

Smaller firms might find that a significant challenge is for them to raise their profile sufficiently within a large, crowded and heavily competed market. Whereas other firms may have decided to ignore their broader profile and concentrate on developing a few critical relationships. It does not matter what your chosen strategy is — but you must make sure that you understand why you are using a particular tool and be clear about the results that you expect to achieve.

Choose your weapons

Above we learned that each of the marketing communications tools has a specific role depending on what you are trying to achieve. So if you need to raise awareness, then advertising and PR are your weapons of choice. If you are trying to initiate a dialogue or provoke a response, then digital and direct marketing is the tool. If you are trying to prompt or hurry action, usually by price incentives, then this is called sales promotion (but be careful to what it does to your overall "brand").

As you are generally involved in a discussion between a particular surveyor or agent and a specific client we are mostly using the time expensive personal selling tool (see chapter 6 on selling and business). And once you have successfully sold and converted a prospect into a client you then need to nurture and grow that relationship and do the elusive cross-selling thing (see chapter 7 on clientology).

A–Z of promotional or marketing communications tools

Rather than set out the various tools and techniques in the usual order that marketers do and confuse everyone, I thought I would set it out alphabetically — which will make it so much easier when you are faced with a supplier on the end of the phone or a partner with cap in hand for some money from the already overstretched marketing budget.

Advertising

In general, advertising is where you pay a media owner (a newspaper, magazine, poster site, website) to display your message exactly as you wish. There is a challenge in selecting the appropriate media for your particular target audience and then is a further challenge in producing an interesting advert that fits with your brand and strategy yet also has impact and differentiates your offer from others while also looking appropriate in the chosen media. For example, the sort of ad you place in a local free newspaper may look and read differently to one you pay for in an upmarket county magazine aimed at the wealthy. Too often, property ads are either dull or they contain too much information from the advertisers' point of view.

It seems that many property people forget that a really good advert needs a number of things:

- A correct understanding of the target audience (and the context in which they will be reading the medium).
- An attention-grabbing headline focused on the needs of the target audience.
- A short, clear description of the service being advertised (or solution being offered).
- Interesting typography and/or photo or graphic that sets your advert apart while preserving your visual identity and brand values.
- A strong proposition (What are the benefits? Why is your firm uniquely placed to deliver?).
- Your contact details — name (people prefer to have a specific name rather than an anonymous organisation), phone, email, address, website.
- Your logo.

Someone has to have a clear idea of what you are advertising and to whom. Then someone has to translate this thought into good clear copy. Then you need someone who is good with photos and design software to set it out in your housestyle. Then you need to send it to the media who will set it in position.

From a budget point of view, as well as paying for the space for the advert to appear you may also need to pay for photography, copywriting and design/artwork of your advert (unless you have invested in getting some templates produced and training your own staff to modify ads as required). It always makes me smile when I see partners in property businesses arguing over whether or not to spend £200 for an advertising slot and then blow over £500 getting the advert produced and artworked.

There are many different types of advertising options for property firms, so here is a review of some of the most common ones.

- **Google Adwords (pay per click)**

  With Google being the dominant search engine and people becoming more and more sophisticated with their search engine optimisation (see below), many firms pay money to have their website appear in a sponsored link at the top or to the right of the generic search results. Google is very helpful in providing tools for you to assess the level of traffic for each of the key words you might use (eg, Homes for sale in Oxfordshire) and allows you to set your daily limit. There are also other keyword tools, such as Wordtracker or SEOBook. However, you need to spend quite a lot of time selecting and refining your key words and also monitoring the quantity and quality of the enquiries that result. Also, it goes without saying that you need to have a good website landing page for your Google Adword campaigns — otherwise the people arrive at your site via the usual Home Page and quickly bounce away again.

- **Local newspapers**

  Many firms will have a number of their properties on a full or half-page spread in the property section of their local newspapers. Please remember that as well as promoting the properties these ads are also supporting your overall local profile and positioning so they need to confirm to your branding and housestyle guidelines. As with most advertising, you need to have regular appearances (ie, multiple insertions) over a period of time for your name to become familiar and trusted. As well as display advertisements, you may sometimes have entries in directories and special features. Some firms have done great deals with just having regular "ears" (the top right or left-hand corner of the main news, television or sports pages). Be careful, as many local newspapers will insist on advertorial ie, a "special promotion" where an advertisement is made to look like a feature article and this could undermine your credibility. Some firms have used a sponsored regular column on property issues to good effect to build local profile and generate leads.

- **Online advertising**

  This is a fast-developing area where it is almost impossible to keep up. Just as one technique takes off and becomes effective, another emerges to eclipse it. I can give you some broad guidelines but I suggest that you contact a real online expert before committing hard cash to this medium.

- Banner ads (55% of all web ads, 96% of all internet ad awareness) are linked to key words in search engines. Click thru rates can be as low as 4%. See the information on Google Adwords above. Increasingly people are using rich media banner ads containing videos.
- Pop ups transfer users to sites, games and competitions but many browsers and security systems will suppress pop ups.
- Superstitials/interstitials appear during page downloads. Again, many browsers will suppress these for security reasons.
- Micro sites are product/promotion-specific and are often run as joint promotions with other advertisers.
- Email includes viral campaigns and loyalty/relationship programmes.
- Portals where you have a free or paid for link from a key site to your own.

- **Posters and hoardings**
  While some property firms have made effective use of posters in train and bus stations (see transport advertising below), relatively few have experimented with the large 48-sheet posters that we see at particularly busy roadsides or road junctions. Those who are working on large residential and commercial developments have made really good use of the hoardings and I have seen some inspired work here — although often it is the developers rather than the agents who have seized the initiative.

- **Regional and national newspapers**
  This is where things become really expensive. A half page in a decent regional or national paper is going to cost you several thousand pounds. The placement of your advert is important too. Often, people will just flick though double-page spreads of adverts. It is good if your half-page advert appears solus among some of the more important editorial material — but you will have to pay for this privilege.

- **Trade magazine adverts**
  Most firms will have adverts appearing in *Estates Gazette*, *Property Week* and other property specialist magazines. The challenge here is that your audience is slightly different — it is usually agents at other firms. And your adverts are typically paid for by your

clients to promote their properties. So how do you balance the needs of your client with the needs of promoting your firm? How do you differentiate?

- **Transport adverts**
  These have become increasingly popular as initiatives such as congestion charges and high fuel prices have forced people off the roads and onto buses, tubes, trams and trains. Most of these poster sites are relatively small and need to have a simple message with an immediate impact. On the London tubes there are now multimedia sites on the escalators — which are highly effective but also rather expensive — both in terms of their production and their space costs. Some of the transport ads can carry more content as the passengers have little else to look at during their journey — so sometimes you can have quite detailed messages. Many property firms have "adopted" particular train stations so that their firm name appears on a board right beside the train station name board. Douglas and Gordon, at Putney in South West London, is a good example.

Blogs

Blog stands for Web Log. It is similar to an informal, open diary or commentary on developments in the marketplace and/or a place for the firm to provide its view on key market developments. As well as providing a reason for clients/prospects/referrers to return to the website on a regular basis (they do this by signing up to see your updated blog entries via an RSS Feed — see below), the regular blog entries help the website remain updated and nearer the top of the search engine lists. So blogs are a key tool in your search engine optimisation.

This is a particularly good tool to use if you have a niche strategy. For example, providing regular updates and information about developments in the rural/agricultural or retail property sectors. It is also a valuable tool of you are pursuing a "rock star" media relations approach to raise the profile of some of your specialists. I have seen good use by property firms of blogs in areas such as green buildings, sustainability, retail parks and in urban development.

The trouble with blogs is that you need to ensure that you obtain regular entries. This means at least twice a month. If you have a surveyor or agent who can write well, then you are lucky but it is more likely that you will need a PR consultant, ghostwriter or an in-house marketing person to prompt the relevant surveyor or agent to identify and then comment on some topical issue and have the PR or in-house marketer draft some copy that the surveyor/agent checks before it is posted to the site.

For an introduction to blogging take a look at the main blogging sites *www.blogger.com* or *www.blogscene.co.uk*.

Boards

I am sure that I do not need to say much about how important the quality and number of sales and letting boards are to the reputation and profile of a local estate agency. Yet recently there have been some rather innovative developments. Like using colour photos of cute animals such as penguins. Or using words such as "Success" rather than "Under offer" or "Sold".

But as boards are a special form of advertising you need to take special care. Your firm's logo/name needs to be clear, which is not so easy if your firm has a really long name. You need to ensure that the telephone number is easily seen from a considerable distance. Ideally, you would like your website address to appear. Of course, the danger is that you try to cram so much on a board that ultimately it all becomes illegible. So invest some time looking carefully at what your competitors are doing. Then invest in some good design to ensure that you make the most of your firm's identity while doing a good job of promoting the properties and differentiating your boards from those of your competitors.

Brochures and leaflets

Some say that the digital age has killed brochures and leaflets but I am not so sure. There are still many people who prefer a hard copy document and there are numerous situations — for example, in your reception area and meeting rooms, at events, in information packs, at visits — when a hard copy document is still useful. If you are in a large firm with numerous offices and teams a brochure can be an important tool for internal communications too.

Having said this we need to be clear about the purpose of the document. You may still feel that despite your website that you need a short document describing your firm, setting out your credentials and services and introducing your people. Make sure that it is not too inward-looking (how many times does it say "we" — and are the readers likely to be interested?). Also, make sure that the look and feel is in line with your overall brand and reputation strategy. Remember that the copy should be interesting to read so use a professional copywriter if your skills are lacking. While good design is important, use a professional so that it is not all gloss and no content. Finally, professional photography should also be used, as amateur shots with poor lighting are easy to spot.

Another tip is that when you produce your document ask the designers to provide you with a PDF (electronic file) — you can use this to email to people, it is relatively easy to do short print runs with digital printing and it means that you can "test" the document before committing to a larger print run. Many firms choose to produce folders with inserts, which makes it easier to both mix and match the particular content for each person you give it to and also to update the individual sheets when you want to make changes.

Aside from brochures that describe the particular attributes of your firm or a particular team or department, you might want leaflets that provide advice and guidance on particular topics or issues. These are valued by clients when they address real problems where they need information. Good examples might be on some recent piece of legislation (eg, disability access, green status, home improvement packs, changes in SDLT and other tax rates) or on topical issues (eg, enfranchisement, equity release, overseas property ownership). In addition to providing information, they position your firm as knowledgeable experts and provide the relevant contact points.

One word of warning though. Often firms will produce great publications and then make it possible for people to download them free from their website. The point of many publications is to create a dialogue so perhaps request their email address before they obtain these documents from your website. This way you have a method of tracking who is interested in what and can even follow up requests to see if you can provide further assistance.

Another key point is that many surveyors and agents will produce a brochure or leaflet without focusing on what they are trying to achieve with the brochure or even the nature of the target audience. This sometimes means that they end up with a beautiful document and no idea of how to reach the people for whom it is intended. So as you tackle the production of any publication, keep in mind how you intend to use it and distribute it. As with all of these tools, the publication itself is not marketing — and you need an integrated plan of activities (a campaign) to use with it to ensure that you achieve your marketing and sales objectives.

Checklist and actions

- Does your marketing strategy indicate the extent to which you are focusing on awareness raising as opposed to relationship development?

- Do all your advertisements — online and printed — have the same look and feel, conform to the housestyle and convey the right messages?

- Do all your advertisements carry contact information and drive traffic to the right area of your website?

- Are the media you use right for reaching your target audience(s)?

- Do you know which adverts are most effective at generating awareness and the right type of leads for your firm?

- Have you considered a blog for keeping your website up to date?

- Are your boards clear and distinctive?

- Do you have print and PDF versions of a document that describe your firm and its services and the value you bring to clients accurately?

Charity

This one needs some care. Too often, firms confuse charitable donations with marketing and PR tools.

If you are making a large donation to a charity, or if a number of your people are dedicating their time to a charitable cause then it seems only fair that — providing the charity remains the key focus — you should be able to use a photo of that situation to try and raise the profile of the charity and your firm in the local press. However, if you are looking for PR alone, then it will be easy to detect whether your commitment to that charity is real or not.

Some firms organise events — such as auctions and quiz nights — in order to entertain clients and contacts while also raising funds for the good cause. This works well although it always helps if there is some clear reason why your firm is associated with that charity. From my perspective, the most mutually successful property firm and charity associations are those that have been in place for some time and people become accustomed to seeing your firm and the charity associated together. Some firms have adopted a local children's or housing project and devoted all their charitable energies to building a strong relationship and making a significant contribution.

Many firms agonise over the hundreds of requests that they receive from national and local charities — partners who have children in school sports teams, members of staff who have a family member stricken with some particular illness, a local charity that is right on your door step and organising an event. Many charities will ask for an "advertisement" in their programme. Firms often forget that while a £50 donation seems fair, it will cost them management time and artwork/production costs (that can often be higher than the original donation) to produce an entry of the appropriate style and size.

Some firms decide to have a separate fund and allow each partner to allocate up to £200 each year on charities of their choice. Other firms decide to nominate just one charity each year and allocate all their raised funds to that charity.

Community events

For many firms there will be a host of country, county, school and other local events where they could take along display panels (which can be produced inexpensively for a few hundred pounds — but please make sure that you keep your firm's name and service clear and the panels relatively simple.

Be clear about what you are trying to achieve. Many firms will go along to these events to show support to others in the community or perhaps to provide an opportunity for past and existing clients to drop by and say hello. Other firms will see it as an opportunity to generate new business leads — but is a family day out the time to engage people in serious conversation when they have impatient children by their side? If you are trying to obtain contact details, then you will need some form of competition or inducement — but then are you sure that the participants are really interested in your services or just keen to win a bottle of champagne?

The other issue is your people. If you have a stand then you need your people to be prepared to devote their time to being available on the stand to talk to passers by and visitors (and other exhibitors). It is sad to see a stand "manned" by people who clearly do not want to be there or are totally engrossed in a conversation amongst themselves. What sort of message about your firm are you giving in these situations?

If you want to support a local event, but do not want to commit the time and resources to having a stand then maybe you can provide support in some other way and make sure that your firm's name is highlighted as a sponsor. At our local schools, there are estate agents who provide all the signage for events — and they do it year after year so that everyone knows that the firm is committed to the local area of the school (see the case study on Chase Buchanan).

Conferences

The property industry is fortunate in that there are numerous conferences both abroad and at home that senior people attend. They are places where it is important to be seen and where you can touch base with a large number of people (clients and fellow surveyors/ agents and potential clients) in a short space of time — so they can be very time efficient. They are valuable for connecting with past and present clients, for obtaining up-to-date market information and for making new contacts.

Ideally, you want to try and have one of your experts on the podium to talk about a subject that is critical to your firm. But you need to invest the time and effort in researching, preparing and rehearsing a good speech and also having the relevant information available as handouts. This makes it easier for people to know who you are and what you are there to talk about. Without a speaker or panel member on the platform you are just another grey suit among the 300 who are hustling for business. Your networking skills — which extend beyond chatting to those people who are closest to you — need to be first rate.

Conferences are expensive. You have to pay for the ticket, the travel and accommodation expenses. Furthermore, your staff are out of the office for sometimes quite long periods of time. Too many firms fail to prepare proper plans for their conference attendance — and there is a long lead time to determine what you hope to achieve, who should go, who you particularly want to meet, what key topics you will be talking about and how you intend to follow up any contacts made.

Copywriting

It has long been an issue for me that while the property industry is happy to pay significant sums of money to professional designers and photographers, they seem reluctant to invest even a fraction of that money on good copy. The result is that most copy is dull and vague.

When you produce a brief for a brochure or a website, you should think carefully about the key messages that you wish to convey and the style, tone and personality of the way in which those messages are conveyed. Back to brand!

We can't all be good at everything and writing in an entertaining way with authority and persuasion is a real skill. Professional copywriters (whether in PR firms or as independents) do not cost a huge amount of money and are seriously underused in the property industry. Or maybe there is someone in your firm who has a skill in this area and they can be persuaded to help others?

Design

If you have a strong brand with a clear identity and guidelines for how all your various materials (brochures, advertisements, boards, exhibition panels, letters, emails) should appear then you will not have to keep spending money each time you produce a new document.

My advice would be to select a suitable designer and ask them to review all your materials and deliver a framework for everything in the future. Ask them to produce some templates and an image library that your own people can use.

Also remember that designers have different specialist skills too — some specialise in traditional printed materials, some in websites and e-marketing campaigns and others in advertising. So choose the right designer for the job. Also have a written brief and get an estimate of the costs in advance. If you want your designer to manage any printing or finishing processes, then understand that you will pay a premium for these services.

Checklist and actions

- Is there a policy explaining how and what you donate to charity?

- Do you have nominated charities each year?

- Do you use your charitable activities appropriately in your PR campaigns?

- What community events do you participate in?

- Which events are most effective at raising your profile and which at generating business?

- Are your people trained to be effective ambassadors for your firm at events?

- Which conferences do you attend? Have you set out what you expect to achieve at each event? And developed plans? And thought how you will follow up?

- How good is the copy in your marketing materials — does it convey the right personality and tone?

- Do you have a good designer who has provided templates for items that you use regularly?

Direct mail

Direct mail used to be about producing lots of letters or invitations for those on your database and directing all your admin staff to the fabulous task of inserting letters into envelopes.

Many firms use direct mail not so much for generating new business but, instead, for keeping in touch and front of mind of their existing clients, referrers and contacts by sending out newsletters or regular updates.

Well, with the advent of technology the old paper mailshots have mostly died out. They have been replaced with an email equivalent. In some ways this is good — it is faster and cheaper to get hundreds of emails sent to people. Some of the software you can use will even tell you exactly how many emails arrived, how many were opened and how many were forwarded. They will even manage your opt-in permissions and integrate with your website lists and main database.

You can use your web-tracking software to see how many people clicked through from the email to a particular area of your website.

However, people receive so many emails that it will be hard for yours to get their attention. And what if they check their emails on their Blackberries or other portable devices only? So be clear about the proposition and the message as well as the way in which this is presented. Be aware that many firms will have email filters and browsers that suppress fancy graphics and images.

Have a read of the section above on databases and also on the section in chapter 4 on segmentation, as having a good quality list of the right target people is fundamental for effective direct mail.

Domain names

If you have a website then you probably already have a domain name that is suitable for your firm. But you may need others — for example, with .co.uk as well as .com. You might want domain names that help you achieve high rankings in the search engines or that provide a separate area for a particular team or branded service that you provide. You should also be careful that if you send a lot of email shots you ought to have a different domain name to your regular email address as you risk having your normal emails blocked along with your e-marketing.

There was an example of an unfortunate incident where the .com for one of my clients that had the .co.uk was owned by a porn site. Their clients were not amused when they went to the wrong domain. Shortly, there will be lots of new suffixes available — we already have .co.uk and .com but they are planning to introduce hundreds more, such as town names such as .Lon (London).

Email

Take a look at the section on direct mail. While email has many strengths, it also comes with some pitfalls.

Entertaining

The property industry is full of party people — any excuse to get together for a drink and a chat! So it is easy to think about all manner of ways to entertain existing clients, referrers and intermediaries and potential clients. Here is a major source of arguments at partner meetings — because some partners like to spend money on expensive events while others simply do not get it and think it is a waste of money.

I have seen all manner of ways to entertain clients — lunches and dinners, going along to the rugby (or football), playing golf, family fun days, jazz in the park evenings, quiz nights, off-roading, quad biking, softball and cricket matches, casino nights, horse racing, clay pigeon shooting, sailing, awards dinners, receptions at key conferences, local community events, arts and music events, garden parties, paint balling, sponsored productions at the local theatre, wine and whisky tasting, personal shopper evenings, cooking events, fishing ..., I am getting exhausted just thinking about all the possibilities. However, people are more likely to guard their precious time so you may need to do more than simply organise a few drinks at an interesting venue to get the right people to come along. Many firms will focus on events where there are things that they want to do rather than what their clients want to do. Even when you have organised an exciting event, have you devoted sufficient time to deciding exactly what it is that you hope to achieve, carefully targeting the guests and inviting them with enough notice and properly briefing the home team on what you are trying to achieve? Is there a clear plan for how you intend to follow up after the entertaining?

Entertaining is a good way to strengthen relationships with existing clients (this is discussed further in chapter 7 on clientology) but needs to be done in a structured way — otherwise the same people come along to the same events every year. And no new business! And with the economy as it is at the moment, you need to guard against being seen to be too generous or flamboyant. We should question effectiveness in developing new business unless it is part of a carefully constructed sales strategy (see chapter 6 on selling).

Exhibitions

I touched on this in the community events section above. There are various types of exhibitions you can attend — major property industry exhibitions (such as MIPIM), business and commercial exhibitions, those associated with county shows and agricultural events, those associated with particular sectors (eg, retail and education), for particular territories at home and abroad and those on particular topics both within the property industry (eg, business parks, green architecture) and those outside (eg, health and safety, human resources).

Then most commercial conference organisers will have some sort of exhibition running in parallel.

Ideally, you will have identified the appropriate events as part of your overall marketing planning. However, if you are considering attendance at an exhibition in an ad-hoc and reactive way, then it probably is not for you.

Even if you find an exhibition that fits well with your overall marketing strategy and appears to attract your target audience you have to think about what you would do on your stand if you go along. You can spend a lot of money building a fancy stand with huge graphic panels and built-in lighting and cabinets to store brochures and refreshments, or you can invest in a relatively inexpensive portable, pop up stand. Then you need to be clear about what you are promoting and why. Also you must ensure that your staff are properly trained to engage the right people in conversation, ask the best questions to qualify appropriate prospects and obtain their contact details and permission to contact them after the exhibition.

Checklist and actions

- Are you still using paper mailshots or have you converted to using email (with the appropriate software to comply with the regulations, manage opt-outs and measure response)?

- Are you confident that all your direct mail generates the appropriate response from the target audience?

- Which of your mailings are designed to keep existing clients informed as opposed to generating new business?

- Do you have the right domain names registered?

- What is your entertaining policy? How do you measure effectiveness?

- Do you have an annual events calendar? Is it communicated to the right people in your firm?

- Do you have a portable exhibition stand — and the right materials to distribute at the different types of event?

Extranets

Private websites — where users have to enter a username and password to gain access — are more likely to be part of your service offering. You may allow your clients to access information about the status of their transaction(s), use their property terrier, review valuable market knowledge and information or use specific applications, such as planning or design tools. Clients seem very keen these days on using collaborative work spaces where all the documents are shared and where there is a private messenger facility. So often you will be using your marketing communications to alert clients and prospects to the extranet-based services that you provide to your clients.

Facebook and social networks

Members of the older generation will struggle with these tools unless they have teenage kids. I started using LinkedIn several years ago — it is a social network for business and professional people with a bias towards those working in the technology sector. In effect, you post up a profile — photo, career history, current projects etc, indicate the sectors and services you provide and recommend people you have worked with (and hopefully they will recommend you too) and then link to your various clients and contacts. It is a good way to stay in touch with people and helpful if you need to recommend various experts to your clients and contacts. Good networkers — who like technology — find it a valuable tool.

Then there are the more popular social sites, such as Facebook. A similar concept — you have a profile and can advise people what you are up to and build a book of contacts. But there is a much more social and informal element here. And therein lies the problem. While there are numerous firms who are dabbling in this arena, it is often frowned upon to use the social networks for business purposes. Nevertheless, some of the larger firms have found the social networks helpful for graduate recruitment programmes.

Again, as this is such a fast-moving area of marketing, there is little help from the marketing experts. But there are some snippets that might help. For example, online communities have been classified into four types (Chaffey and Durlacher):

- Purpose (same goal — such as lobbying for lower stamp duty).
- Position (circumstances — for example, owners of listed buildings).

- Interest (hobby or pastime — eg, small-time property investment).
- Profession (eg, those involved in surveying).

Muniz and O'Guinn indicate that these online communities share three core components and five characteristics:

- Their model of communication (visitor to visitor — there is no central control).
- They create an identity for members.
- Relationships (even friendships) develop among members.
- Common and/or specialised language.
- Methods to regulate and control behaviour.

This area is very exciting and new, but littered with pitfalls and dangers. So please tread extremely carefully.

Giveaways

These are handy as "thank you" gifts to existing clients that also keep your firm's name front of desk/mind and some firms use them to distribute at exhibitions and other events. Obviously, your firm will have some pens and notepads for your meeting rooms, and you probably also have golfing umbrellas.

Agents will often have keyrings. Commercial surveyors and agents will have all manner of sports kit — polo shirts, rugby shirts, caps — for their teams and friends. Then there are an array of desk top items — pen holders (I once saw a great design for one of these in the shape of a pair of green wellies for one of my agricultural clients), mousemats (particularly helpful if they contain important reference information and/or flowcharts for complex processes), paperweights, calendars, calculators, sticky dispensers, highlighters.

Within the property industry I have seen some extraordinary items. High-quality shovels and spades to commemorate ground-breaking celebrations. Silver wine stoppers and champagne kits. Logos on beach towels for conferences taking place in seaside resorts. Jars of chutney for a particularly innovative direct mail campaign. Beautiful limited edition prints for special development locations.

These items do not have to be expensive and they can be valued by clients — especially if they are useful, everyday items. But try to be creative in sourcing something special — there are numerous online providers who can obtain unusual items when requested. Moreover, make sure that in addition to your name and contact details they carry any important message for your strategic marketing campaign.

Hospitality

See the section on entertaining above.

Intermediaries and referrers

It is worth considering your intermediaries and referrers separately. While it is OK to invite them along to the general entertaining, hospitality and seminar events that you organise for clients and prospects, you should also have a separate programme — mostly of relationship development activities (the chapters on selling and clientology will provide some ideas here).

For those intermediaries and referrers where you have a special or strong relationship, you may want to develop joint marketing activities. You will probably also want to organise structured meetings to explore all the ways in which you can work together — and build a plan. Once you have worked together a bit, I encourage my clients to have a detailed agenda for such discussions so that you have a clear understanding of your planned collaboration. Here's an example agenda:

Proposed structure for referrer meetings

1. Introductions
   a. Your team.
   b. Their team.

2. Aims and agenda

3. Overview of your firm — including current client profile and targets
   a. The various departments, offices and services.
   b. Recent developments and planned initiatives.
   c. What you are trying to achieve and your target audience.

4. Overview of their firm — including current client profile and targets
   a. The various departments, offices and services.

     b.   Recent developments and planned initiatives.

     c.   What they are trying to achieve and their target audience.

5.   Areas of past cross-referral (levels of work and nature of clients)
     a.   From their firm to your firm.
     b.   From your firm to their firm.

6.   Areas where further cross-referrals/collaboration might be possible:
     a.   Services and expertise.
     b.   Markets (domestic and international).
     c.   Specific clients.

7.   Potential joint marketing initiatives
    Internal
     a.   Further introductions between partners/departments.
     b.   Further introductions between more junior staff (below partner level).
     c.   Internal training sessions.
     d.   Work experience exchanges/secondments.
    External
     e.   Articles in each others' newsletters.
     f.   Material/links on each others' websites.
     g.   Collaborative articles/media coverage.
     h.   Invitations to/attendance at each others' specific team/client entertaining.
     i.   Joint client receptions and/or seminars — joint campaigns (possible topics?).

8.   Managing conflicts and relationships with other firms

9.   Agreed actions going forward
     a.   Responsibilities and timeframes.
     b.   Cost sharing.

10. Any other business

There is more information about developing relationships with referrers and intermediaries in chapter 7 on clientology.

Internal communications

Sometimes, people are so focused on alerting clients and contacts to new developments, services and people that they forget to inform the army of ambassadors among their own staff. In a small firm this is rarely an issue. But as firms grow they need to have specific communications programmes developed for their staff. You do not want your own staff learning about developments from the media or their clients first do you?

Many firms will have annual staff days where a series of partners will stand up and talk about the firm's strategy and developments in the market. Some firms will have more regular team meetings. Hopefully, firms will regularly email information about new clients, new services, the latest newsletters or key events to keep everyone informed. Many firms will have intranets where staff can find information for themselves about what is happening.

In chapter 8 on people, we will see that good internal communications is vitally important to instil a sense of involvement and commitment among your staff — and crucial for attracting and retaining quality people. Make sure that you devote some of your time to considering how to communicate effectively with this very important audience that often goes neglected.

Intranets

This is a private website for people within your firm. As well as being valuable for providing information about the firm's strategy and progress, its clients and its people, it is also a valuable way to help teams share knowledge and manage client relationships.

Hopefully, you will have sections on your intranet containing information about marketing and business development as well as the myriad other things that staff need to know. An intranet is a vital tool within your internal communications strategy.

Checklist and actions

- Have you explored how you can use extranets to provide additional value and services to your clients and lock them into your firm?

- Do the younger members of your firm participate in the appropriate social networks?

- Are you involved in any appropriate online communities?

- Do you have a stock of appropriate giveaways and policies on how they are to be used and by whom?

- Is there a list of your most important referrers and intermediaries?

- Is there a structured plan for how you intend to develop those key referrer relationships?

- Do you have a good internal communications strategy and numerous activities to keep all staff informed and involved?

- What does your staff think of your internal communications?

- How well do you use your intranet to support good internal communications and knowledge sharing?

Media relations

See the section on profile, publicity and PR.

MIPIM

It is unlikely that smaller firms will consider exhibiting at this major annual event in Cannes — but it is such a highlight of the property world's marketing that it must have a mention. Smaller firms might consider sending along a couple of partners as delegates to the conference. But if you think that four or five days in non-stop, alcohol-fuelled networking among nearly 30,000 property professionals is good fun — then think again!

Mobile marketing — SMS

I have seen a few agents use SMS as part of their case management programme — alerting buyers and sellers to key stages in their transactions. There were a few experiments using proximity SMS — where messages appear on mobile phones that were close to automated advertising panels or hoardings. There are also a few examples of advertising campaigns where competitions, collaborations or special offers are promoted and clients are encouraged to text a particular word to a number in order to obtain information. The upside of this technology is that it is measurable and has a direct and interactive relationship with the target individuals. The downside is that it is new, can be expensive to set up and some people find commercial SMS rather intrusive.

Networking

Where would the property industry be without networking? Some think of it merely as cruising around a room full of people with a glass in your hand. Others consider it a military operation with detailed preparations and briefings, allocated targets and carefully scheduled follow-up campaigns.

We network at industry events and client socials. We network among our friends and neighbours. But are we just chatting or are we pursuing a proper marketing communications strategy or a researched sales strategy? Some firms invest in networking training or encourage senior members of the firm to take youngsters to events.

Newsletters and magazines

Please refer to the section on brochures and leaflets as many of the points apply here.

Increasingly, firms produce regular newsletters and magazines (some are glossy productions using contract publishers and well-known journalists) in order to have something to remind clients on a regular basis about their services. Larger firms will produce publications that form an important part of their service in keeping clients up to date with developments in their field or to provide information.

Please be aware that clients are inundated with such newsletters and magazines and that there is some extremely high-quality information available for free from the internet, which they can obtain easily through RSS feeds and such like.

Sometimes newsletters and magazines focus on the firm and its developments. These can be a bit too inward-looking and have little of real value to clients. They are valuable when your experts in a particular subject have taken the time and trouble to do a lot of research into a fast-moving or complex subject and set out exactly the sort of information that clients find of value.

I recommend that clients look at all the newsletters produced in their sector, region or area of expertise to see what already exists. It might be possible to contribute to or collaborate with these in some way rather than invest the time and money in writing, producing and distributing your own version. Some firms buy "canned" newsletters, where they simply apply their own logo and perhaps one lead article.

Online marketing mix

Websites, emails, intranets and extranets, online advertising — all these together make up your online marketing mix. There are sections on these separate elements but it is worth thinking about your overall strategy for online marketing as part of your communications planning.

Cartellieri offered a checklist for developing suitable goals:

- Delivering content. (Are you simply trying to provide information and/or convey your expertise?)

- Enabling transactions. (Do you want people to conduct business online or maybe pay their bills?)

- Shaping attitudes. (Are you trying to position your firm or experts in a particular place within the market?)

- Soliciting response. (Do you want them to sign up to newsletters or to contact someone?)

- Improving retention. (As part of your relationship management programme, is your online presence mostly to provide additional services to your clients and "lock them in" to you?)

Checklist and actions

- If you attend MIPIM is there a plan about what needs to be done before, during and after the conference?

- Have you explored the potential for using SMS and other mobile marketing tools?

- Are your people encouraged to attend the right (targeted) networking events?

- Are your people trained in how to make the most of networking events?

- Do you produce a newsletter or magazine for clients and contacts? If so, how effective is it? Have you asked your clients for their views?

- Does your marketing strategy address the appropriate balance of online and traditional communications?

Profile, publicity and PR

There are some confusing terms here. PR usually stands for Public Relations — and this is about managing the flow of information between an organisation and its publics. The "publics" of a property firm include their staff, their potential staff, their clients, their potential clients, fellow property professionals, the local community and the media. Sometimes, people confuse PR with Press Relations, which is concerned with getting your name in newspapers, magazines, radio and television.

The correct term for getting your names in the print and broadcast without advertising is media relations. I have so much to say on this topic that I have written a separate book, *Media Relations in Property*, in conjunction with a leading residential journalist Graham Norwood that is also published by EG Books.

The advantages of media relations are that it is inexpensive (compared with advertising) and has a high degree of credibility. The downsides are that the media (in the form of the editor) has ultimate control over what does and does not appear and that it can take a lot of time and effort to cultivate relationships with journalists and prepare material that is suitable for publication.

There are different ways to think about publicity — there is that which you do about your firm — its successes, people and major developments — we regard this as corporate PR. And publicity about the specific services you promote — this is product PR. Often, property companies are involved in generating publicity about their own clients developments and buildings, so it is easy to become confused about who or what you are promoting — but inevitably you should receive some piggy back advantages when you are publicising your clients' projects.

Ideally, you should have a structured media relations plan that integrates with your other marketing communications, selling and client relationship activities. Within your PR plan you may have PR events (one-off events for a specific purpose such as a new partner or a branch opening), PR campaigns (activity with specific start and end dates involving several events or techniques but all supporting a particular theme or message) and a PR programme (ongoing awareness building and maintaining using multiple techniques). There are numerous reactive and proactive techniques involved from answering journalist enquiries, submitting directory entries, organising press interviews and visits and through to preparing news releases and articles.

Another aspect of publicity is to think about what key messages you wish to convey about your organisation and also those perceptions that already exist. Some firms undertake research to find out how members of the media and their clients perceive them and plot these attributes on a chart so that they can determine what messages to stress.

Figure 5.4: Research and plot key dimensions on a multi-attribute scale

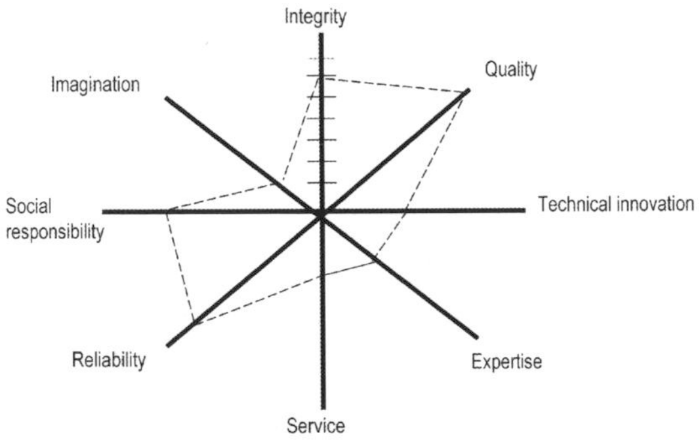

Source: Kim Tasso

It sometimes helps if you produce similar charts for your main competitors, as it shows where you may have some competitive advantage and/or how you might tailor your marketing to stress the valued differences.

If you have experts in particular fields you might explore what we call a "rock star" approach to media relations — profiling your experts, training them in front of camera or interview techniques, arming them with pithy sound bites and encouraging them to contribute comments and opinions about all manner of things in their area of expertise.

Podcasts

Rather than writing a long document about some particular specialist topic or updating clients about new developments in writing or on a CD you might consider producing a podcast that can be downloaded from your website or one of the broadcast sites onto an iPod or similar MP3 player. In the olden days, some property firms recorded their annual market overviews and produced audio cassettes for people to listen to in their cars. It is the same concept but using different technology.

It does not have to be expensive either — digital recorders are easy to get hold of and there are some useful software packages to help you edit the recording and incorporate introductory or background music. However, when I listen to my iPod it is usually when I am in the gym or relaxing on a train. Do I want to listen to someone give me a fairly dull update of developments in the property investment market?

Presentations

Presentations can be informal — and for just a few people from a key organisation. Or they can be more formal for a seminar of 50 people or a conference of 500. They can be used to raise awareness of your firm, its services or its experts or to persuade a particular organisation to select your firm for a particular project as part of a tendering exercise. Increasingly, it is possible to use technology to develop electronic presentations that can be sent via emails or delivered real-time through a webinar.

There is more information about the way in which to develop and structure your presentation and the skills you need to deliver a presentation confidently in chapter 6 on selling.

Research

Research is a very broad topic. Rather than being concerned with research that you use to inform your business, marketing and client decisions we are talking about using research as a component of your marketing publications — whether leaflets, newsletters or websites — to attract and inform existing and potential clients as well as to position you as an expert in a particular area.

The larger property firms have large departments of professional researchers who keep track of vacancy and occupancy rates, rent levels, significant commercial transactions, property price and investment indices and developments in key markets such as retail, industrial, care homes and education. These are published on a quarterly or annual basis. There are some particularly good examples where these surveys become part of the industry's knowledge base eg, the Drivers Jonas Crane Survey.

In public relations this is aligned to issues-based marketing. You pick a topic, undertake or pay for the research and use the results as a platform for generating media coverage, mailing targets and presenting seminars and conferences. You can conduct the research yourself or pay for professional researchers to do it for you or even collaborate with a local business school or college. Increasingly, online research companies (eg, *www.onepoll.com*) allow you to set up and run a research exercise for relatively little investment and then you can use the research results for a media campaign.

RSS feeds

RSS (Really Simple Syndication) is a web feed used to publish frequently updated works, such as blogs and news headlines, in a standard format. An RSS document (which is called a web feed or a channel) includes the text plus information about the dates and authorship. It allows readers who want to subscribe to timely updates from favoured websites or to combine feeds from many sites into one place.

It is a useful tool for research purposes — to keep track of updates on a variety of websites focused on a particular subject and it is also beneficial if you offer an RSS feed from your website news section or from the blog so that people can easily see when new content is posted.

Sales promotions

In technical terms, sales promotions are special offers that encourage people to purchase now rather than later. Often they involve some sort of price inducement — while there are not too many "buy one, get one free" offers in the property world there are some recent examples of agents offering to do house sales and purchases at cut prices within a particular period to stimulate demand.

Price promotions are dangerous. Not only do they bite into profit margins but they can change forever the market expectation of a fair price when the promotion is over. What's more, they can do damage to your overall brand and reputation — so if you are known as the Rolls Royce service or the Gold Standard, think carefully before you undermine your position and offering in the market with a low price. Also, if you try to offer something for "free", then people will be naturally suspicious.

Competitions — if you can think of an attractive idea — are a good way to encourage people to visit your website or your offices or a stand at an event. But there are some legal rules if you get involved in people buying tickets or lotteries so please check these out. Many firms will have an assortment of giveaways that are used as part of their more general marketing and client relationship programmes.

Sales visits

See chapter 6, which is devoted entirely to professional selling.

Checklist and actions

- Do you have an appropriate media relations adviser?

- Do you know the relevant journalists at the appropriate national, regional, local, property, business and specialist media?

- Do you produce news releases and articles on your firm, people, services, clients and transactions?

- Are you clear about the key messages you wish to convey to different audiences and media?

- Are your people trained in how to deliver great presentations?

- Do you have standard presentation materials about your firm and its services?

- Do your people know how to use RSS feeds to keep up to date with specialist topics and as part of their research activities?

- Are you careful when you use price promotions not to undermine your overall positioning in the market?

- Are you confident that your people make the best use of all sales visits?

Search Engine Optimisation (SEO)

When clients use search engines such as Google or Yahoo to find out information about property firms or services you ideally want your website to appear towards the top of the first page of results. To achieve this, your website needs to be optimised. The alternative is that you pay for online advertising — such as with Google Adwords — so that your website appears at the top of the page or to the right as a sponsored link.

The "rules" for optimising change frequently and this makes it rather difficult to keep up. A good web developer will know how to build a well-optimised site although you can hire the services of specialist search engine optimisation experts. Such experts usually operate on the basis of a number of key words or phrases and you have to pay them on a monthly retainer basis.

Whether you plan to do the optimisation work yourself or to use an external expert, you should be familiar with the basics and there is an excellent beginner's guide to this complex subject at: *http://google webmastercentral.blogspot.com/2008/11/googles-seo-starter-guide.html*. The present guidelines indicate that you should:

- Design unique, accurate page titles.
- Make use of the description meta tags and heading tags in your HTML code.
- Structure your URLs (web address) carefully.
- Make sure you have accurate descriptions of any images you use.

- Have a sitemap.
- Use key words within good quality content as hyperlinks to other parts of the site.
- Include key words as part of the navigation (ie, within the menu system — a breadcrumb trail would look like this: property — landlord and tenant — dilapidation disputes).
- Update the content regularly (a blog might be one way to do this).
- Encourage other people to have links to your site, particularly those from "authority" sites such as wikipedia.

Seminars

Seminars are a proven tool to raise awareness of some special expertise that a firm has, to showcase your people, and to initiate a dialogue with the right audience. It is also a good opportunity for those who do not know your firm to "try before they buy" and become familiar with you in a low-key way. Seminars are also highly valued by existing clients as part of the service you provide to keep them up to date and perhaps even for free CPE (Continuing Professional Education) points. Clients often value the opportunity to network with their peers and people in your firm that they may not have previously met. So make sure from the outset that you are clear in terms of what you expect to achieve with the seminar and who, ideally, you want to attend — the mix of existing clients, targets and referrers.

However, seminars are used a lot by property firms and so you need to do something special to attract the proper number of the right quality people. Also, people are notorious for accepting invitations but then not turning up (as a rule of thumb, expect around 30% of acceptances not to show up on the day).

Organising a seminar takes a huge amount of work. You need to have a good list of people to invite. You need a suitable topic and excellent speakers (ideally some from outside your firm to add interest). While it is a great opportunity to get people into your offices, if you do not have adequate space or facilities, then you will need to find a venue that fits with your usual style and/or provides an added attraction to the intended audience. Then there is the cost of the catering.

Then you need to ensure that your speakers put together interesting and relevant presentations and that they work together to ensure that the different talks integrate well — without overlaps or repetition. Getting surveyors and agents to rehearse in advance is one of those perennial problems. There probably needs to be a pack for each delegate and a carefully constructed plan of who from your firm will aim to speak to the various delegates.

Smart firms will think about how they can repeat some or all of the sessions on different days and at different times. You can, of course, recycle a lot of your seminar material to be used in your PR and on your website and within other marketing campaigns. But you also need a good follow-up plan — too many people sigh with relief and sign off when the seminar is over — when this is really the time for the hard work to begin (Chapter 6 on selling may help).

Finally, make sure that your seminars are part of an integrated programme of direct marketing, PR and selling activity — the sales cycle for most property firms is usually several months.

Social networking

See the section on Facebook and social networks.

Sponsorship

This is where you pay money to have your firm's name associated with some other organisation or event. Some of the larger firms sponsor major international sports events and you see their banners at the edge of the pitches and in the programmes. Those larger firms also support the arts with sponsorship of major exhibitions at the larger galleries. But this does not necessarily mean a big cheque.

Local community organisations (eg, charities, schools, sports facilities, theatres, clubs, groups etc) and events (eg, summer fairs, open days) often do not charge much for you to be a headline sponsor — acknowledged in all the advance publicity and programmes on the day — and it is easy for you to have some form of presence at these events (see exhibitions above).

Some firms will sponsor their own staff who are participating in charitable, hobby or local interest activities and provide them with clothing that is smothered in their logo. (See the section on charity above.) They then place photos of these adventures on their website in the news section or in the area dedicated to Corporate Social Responsibility. Other firms will have a long-term association with a particular organisation or charity that is within their community or sits well with their chosen markets or areas of specialisation.

I would guard against the knee-jerk reactive types of sponsorship where you are asked for support. As part of your marketing planning you should look at the sorts of groups and events that fit in well with your overall aims and targeting and make the first approach. Negotiate to ensure that you obtain all the relevant benefits that you want from your sponsorship money — whether this is in terms of branding, collaborative marketing, publicity, information in packs, free places, mailing lists and so on.

We used to have a rule of thumb that indicated for each pound of sponsorship money you would need to spend a further pound with supporting activities (whether advertising, publications, events, entertaining, time) to make it work for you.

Virtual worlds

There are a few professional firms (but not too many in the property sector) who have started to experiment with virtual worlds such as Second Life. However, most of them are larger firms who are focusing on graduate recruitment programmes. However, if you have a good example of marketing within a virtual world, I would love to hear from you.

Webinar

This is the new technology alternative to all the time, money and expense of organising a real seminar (to which it is notoriously difficult to get the right people to attend). In essence, you use some special software on your PC (many people use Microsoft Sharepoint, others use systems such as Gotomeeting) that allows people from all over the world to log into a user account and see your Powerpoint or other presentation materials on their PC but which also allows them to raise questions with the moderator (the person controlling the interaction), the speaker or with other participants. They usually have to also dial in to a particular teleconferencing service so that they can hear the words on their speaker phone although some of the newer teleconferencing systems allow them to see live video footage of the presenter and hear their voice through their PCs.

The upside of this technology is that it is relatively easy to produce and present a seminar. Also, as there is no travel involved, people are more likely to book the 30 minutes or hour to join the webinar in the comfort of their own offices. Also, if your speakers get stage fright when standing in front of the audience it can be easier for them to talk at a screen completely oblivious to the numerous people who have joined the event.

The downside is that the technology can let you down — either if the connections at your end go down or if your participants are not familiar with the relevant joining instructions. Another major downside is that you miss out all of the important face-to-face interaction — you do not know how your "audience" is reacting to what you are saying. However, as a service to existing clients or to gather a number of people in remote or international locations it is a real God send.

Websites

Even the smallest property firm is likely to now have a website. More likely, it will have a number of areas around which you navigate to provide profiles of your people, the properties that are being marketed, some background about the firm, descriptions of the various services being offered and maybe more up-to-date information about developments in one or two specific markets.

Sadly, too many property firms have rather dull sites that are little more than electronic versions of their brochures, with the copy being dull and almost entirely inward facing ("We have been established for nearly 100 years and are one of Dullstown's leading ... blah blah blah"). The content is rarely updated and there is little reason for people to return to the site once they have obtained the contact details they were seeking.

If you want to be more ambitious and have a highly interactive website with lots of visitor facilities and up-to-date information so that it is "sticky", then you will have to invest in having someone prepare the regular new copy and someone else to post it onto the site using a CMS (Content Management System).

The marketing academics have not really caught up with all the latest developments in the world of web but there are a few helpful frameworks around to structure your thinking. When designing your website, you should consider four key aspects (Karayanni and Baltas) — interactivity, navigability, multimedia design and content (your firm and your clients). There is also a helpful 7Cs best practice framework (Rayport and Jaworski):

- **Context**
  Is this your firm's site for visitors with a specific interest, providing in depth information about a particular topic?

- **Content**
  Is the quality of information up to scratch? It is up to date and accurate? Is it likely that visitors will need to return regularly to see updates? Is there a consistent style?

- **Community**
  Do visitors feel part of a community? Do they have sufficient in common? Is there a way for them to interact with you and with each other?

- **Customisation**
  Can visitors tailor the experience and information to their particular needs? Can they arrange to obtain updates on their areas of interest? Can they opt out of information that is of no interest to them?

- **Communication**
  How do you promote interaction? Are your contact details clear — telephone, address and email? Have you explored instant messaging or call centres?

- **Connection**
  Do visitors feel a real connection with the site and its contents and you and other visitors?

- **Commerce**
  How will it support your commercial objectives? Generate leads? Retain clients? Cross-sell services? Form part of your service? Support transactions? Reduce operating costs?

Expert Chaffey advocates looking at standards in the following areas:

- **Site structure**
  Vital for search engine optimisation, but also to make it easy for visitors to get to where they want quickly — no more than three clicks.

- **Navigation**
  Important for both search engine optimisation and ease of use.

- **Copy style**
  Particularly important if you have numerous contributors, so make sure you have adequate procedures and processes for checking material before it is posted to the site.

- **Testing standards**
  Do you rigorously test changes to the site before making them live? Do you test with different PCs, operating systems, speeds, browsers and screens?

- **Corporate branding/graphic design**
  While it is tempting to become creative on the internet, your site needs to look and feel similar to other materials from your firm.

- **Process**
  What processes are in place for approving new material? For deleting old material? For responding to requests and enquiries? For checking the accuracy? For monitoring competing sites? For compliance with legal requirements?

- **Performance**
  How quickly should it load and move between sections? How long must your visitors wait to receive a response?

Many modern websites will have an integrated CMS (Content Management System) that allows people who are not trained in website design (typically using web code which is called HTML or PHP and tools such as Abode Dreamweaver, Flash, Fireworks and Photoshop (for image manipulation)) so that regularly changing sections of your website, such as news, publications, people, can be updated in-house without incurring the cost of your external developer.

It is worth mentioning here the need to have a clear web strategy before parting with good money to a team of web designers. You need to think through how you intend to use your website — is it to be just an electronic calling card or do you want it to take centre stage in all your marketing strategies? Will you be providing applications and services to your clients through extranets (see above)? Will you be using databases to manage large amounts of information? Will you be investing time and money updating complex content or directing traffic to your site through an extensive online advertising or SEO campaign? You need to have a clear strategy — one that integrates with the rest of your marketing plans — and to develop a good brief to explain to the designers and developers what you want to achieve. You will also need to speak to your technology people to determine where and how the website will be hosted and maintained. You should also see the section above on domain names.

Also, you should make sure that your website includes a facility to monitor the number of visitors and the pages that are most and least popular. You can include some simple code on each page of your website to use — for free — a tool such as Google Analytics, which will monitor many types of data and provide helpful charts to show you who is visiting your site and how they arrived there.

Wikipedia

This is a free, multilingual, open content encyclopaedia project operated by the US-based non-profit Wikimedia. It is unusual in that it is written collaboratively, so anyone can contribute information to it although they encourage people to verify the accuracy of each entry. Since its creation in 2001, it has grown rapidly into one of the largest reference websites. It indicates that it has had nearly 700 million visitors a year and 75,000 active contributors working on more than 10,000,000 articles in 260 languages. Many people use it as a source of information about a topic (try looking at the page on surveying) and a few firms have set up pages on their own sites with well-referenced information to which Wikipedia will link.

Word of mouth

Many firms — particularly those that are long established within a particular community — will rely heavily on word of mouth recommendations from their past and present clients, referrers and intermediaries in the market and even their own staff. Broadly speaking, word of mouth is a part of public relations and is difficult to manage in a proactive way. However, there are some frameworks (Bone) that can help you think about how you can support and encourage good word of mouth recommendations:

- Direction — Clients may seek information and/or recommend-ations about your firm before they contact or instruct you (input) or they may tell others how they felt about the service you provided (output).

- Valence — Do clients generally have positive or negative feelings as a result of dealing with your firm and its people?

- Volume — The number of people talking about and hearing about your firm.

A useful exercise here is to ask clients why they came to you at the outset of a transaction or relationship and then, when a transaction is complete, to ask them whether their expectations were met. Client satisfaction surveys (see chapter 7 on clientology) can usefully ask clients to rate their propensity to recommend your firm on a scale of one to nine.

You can provide case studies and testimonials from past satisfied clients in your brochures and on your website and offer lists of "reference sites" to commercial clients. This sort of information is particularly valuable when involved in tendering situations.

Another important strategy is to ensure that the key influencers and opinion formers in your market are properly informed and kept up to date with developments at your firm. This might be an important part of your relationship programme with referrers and intermediaries.

Figure 5.5: Opinion leaders and followers

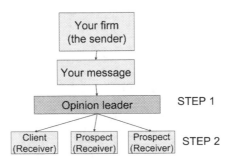

Source: Adapted from Chris Fill

Opinion formers are those designated as knowledgeable about a subject — their advice is credible and they are relatively easy to target (eg, journalists, investment analysts). Opinion leaders are those who reinforce the marketing messages sent, and to whom other receivers look for information and advice. These are much harder to identify and target. The important thing to remember is that whilst you can manage and control the message that you convey to opinion formers and leaders (step one) you have no control over what they, in turn, pass on to the ultimate clients (step two). Furthermore, because each opinion leader can have an impact on many receivers extra care should be taken.

To manage a word of mouth campaign (Ennew, 2000) you should:

- Listen actively.
- Question effectively.
- Be client-oriented as an organisation (this will be dependent on your culture as well as your strategy).
- Consistently deliver on promises — this means ensuring that all members of your firm understand the importance of your quality standards and style of working.
- Help clients in information seeking.
- Manage internal word of mouth (this relates to your culture and internal communications and training programmes).
- Find out what competitors and clients are saying (and you should invest in both formal and informal research programmes so that you really know — there is more on research in chapter 7 on clientology).

Checklist and actions

- Have you checked that your website uses the latest search engine optimisation guidelines?

- Do you have clear objectives for your seminars with detailed plans for what needs to happen before, during and after an event?

- Do you recycle seminar materials in PR, web, direct mail, client relationship and other activities?

- Have you explored whether some of your seminars might be more cost-effectively produced as webinars?

- Are you clear about what and why you sponsor and the return on your investment?

- What is your website strategy? Is it mostly static or do you have a programme of regular updates to add value to existing clients or attract new clients?

- Do you monitor your website traffic regularly?

- Have you explored whether you can contribute to wikipedia in your areas of expertise?

- Do you actively manage word of mouth campaigns?

Better briefs — choosing and using external specialists

How many design, PR and web consultants have I had to reassure because their property client kept changing the brief and then argued about the cost? How many irate partners have I had to sit down and explain that the reason that the cost of their little brochure was 10 times the original quote was because he or she (and the other partners) changed their mind so many times about the size and shape of the document, rewrote the document 23 times and urged the poor supplier to "find better photos" on eight separate occasions?

The answer to these problems is simple. Before you instruct a designer, printer, web developer, PR consultant or any other type of specialist you must write a proper brief explaining what you want to do, and why, and the essential elements and critical success factors, and deadlines, the estimated budget and the decision and approval team.

In the book *Media Relations for Property*, which I co-wrote with Graham Norwood there is a chapter on how to develop a brief, select and manage external designers. The process is similar whether you are appointing PR consultants or advertising experts or any other consultant.

But what *really* works? Measuring effectiveness

There is a famous quote that says: "I know that 50% of my advertising spend is effective — trouble is, I don't know which 50%" — or something to that effect.

A frequent question is about the relative effectiveness of one marketing tool against another. But this misses the point. There are three key things to bear in mind:

1. What are your objectives?

   The only real way to measure effectiveness is to be clear at the outset what you are trying to achieve and to have SMART objectives. This way, the most effective tool is the one that enables you to achieve these objectives most quickly and inexpensively.

2. Measure process as well as results

   You need to adopt a long-term approach with many marketing tools before you start to see results. For example, you may spend several months meeting with and talking to a journalist before you start to see your name mentioned in news items and your articles published. It may then take several months before the readers become familiar with your name and your area of expertise and start to make contact with you. In these types of cases, you need to measure the activity and process taking place as the results take longer to appear.

3. The tools work together

   Often, it is a combination of factors that work together and reinforce each other during the campaign to generate a lead. For example, a client might see your firm's name advertised in directories and on posters before they search for information about you online to find your website. They may then subscribe to some of your newsletters and attend some of your events and receive some direct mail items from you before they request a meeting to talk about their specific needs. So what was it that generated the lead? It was an integrated marketing campaign! Chapter 4 on strategic marketing addresses this in more detail.

Checklist and actions

- Do you prepare a detailed brief each time you commission an external designer, web developer, PR adviser or other expert?

- Who has ultimate responsibility for managing the relationship with these external advisers?

- What is the approval process for any marketing materials?

- How much time do you devote to briefing and supporting external advisers in fulfilling their briefs?

- How well do you measure the effectiveness of all the various marketing activities you use — both separately and together?

- Do you have campaign plans that blend together the various tools over a period of time to achieve your stated objectives?

Case study: Colliers CRE — tactical marketing

Margaret Emmens is the former director and head of marketing at Colliers CRE

Not another brochure ...

Grabbing people's attention in a crowded marketplace is a continual challenge. All your competitors have the same end in view — good client retention and a steady flow of new business. So how do you persuade people that you and your team are the best and most expert people to deal with? Well, in the era of spin and sound bite, we have learned that you have to be very sure that what you are sending out is what your audience wants to hear.

What is becoming clear through our research is that, for all the obvious reasons, a brochure is not necessarily the best solution. This is borne out by the startling statistic found in an academic journal that the average reading age of an adult today is the same as that of an eight-year-old in 1945. Not because people are lazy or totally averse to print, but because there are so many ways in which we get information presented to us, from headline television news to screens and text messages.

Do people read text? Well some do, and some don't. How many of us carefully read the instructions for the new washing machine or the coffee-maker? We might give the instruction manual a cursory look, but we are much more likely to press a few buttons and see what happens next.

What is the best plan of attack?

If brochures are not always appropriate, where do we go from here? The way we believe is to go is more forensic science, less spray and pray. More empathy, less column inches.

Where do you start? It is an old cliché but as with all clichés there is a fundamental truth — you need to stand in your client's shoes and view the world more from their perspective. Find out what clients want, and more importantly will want in the future from you as a business adviser. Where can you help them, how can you add value to their future plans?

Less is more — don't deluge people with information. Rambling e-mails that fill the screen but offer no obvious benefit to the reader are just too much stress and will be regarded as a waste of precious time. Don't make all your contact with a client/prospect arms-length and remote. Research shows that more than four emails without a call will be viewed as contact from just another indiscriminate service provider.

Listen. It sounds basic but so many people miss out on business opportunities because they do not attune their ears to the business signals. Most of us will have, at some point in their working lives, been bored rigid by salesmen who talk at us about their product without giving the potential buyer the opportunity to outline what their real business needs are.

Give a good account of what you have to offer

Have a form of presentation available that that can be tailored to what is required on the day. No death by PowerPoint, but rather more key discussion points. There is nothing worse than being the victim of a long rambling set of slides that don't focus on what your real issues are. One size does not fit all.

At Colliers CRE, we developed a form of Flash presentation that enabled our people to give a short "taster" or a longer pitching presentation from within the same basic file. We found that it was a good idea to analyse the working style of the person you want to get to know. If they are the type of person who likes arguments in a logical sequence, then present to them in that way. Others prefer to get the solution first and then backfill with the process and detail, and if time is short it is vital to know the style that suits best.

Use wit. We are not talking about belly laughs but more about showing a lighter more human perspective on the world. Being approachable and open with people and allowing them to see other aspects of your character can really work. Before deluging a client with information, allow them to self-select.

Have a plan and stick to it

We have learned that enthusiasm and consistency are important. It is so easy to start a programme of client marketing or contact and then give up before it has had time to establish itself. Our experiences have taught us that:

1. There is a real need for mapping the decision makers. Corporate purchases of any type require considerable numbers of people in most organisations and it is a slow burn. You have to maintain and grow all the contacts in a planned and focused way.

2. Customers value activities in which they feel in control. You must find a way of ensuring that the information you supply is timely and valuable and delivered in an appropriate fashion.

3. People differ — some value relationships, some value information, some value a combination of the two. It is a mistake to assume everything is either one or another. "Don't sell me a product, solve my problem" came through loud and clear.

4. Stay in contact, return calls and emails promptly and keep them "on message".

5. You need to have an overview of what your company is about. What is new and what is relevant — make the buyer pleased he bought into you and your team by reinforcing messages about the successes of your company. We run "Know your Company" sessions open to all to help insure we are regularly updated.

6. Demonstrate value — understand the difference between features and benefits.

7. Differentiate yourself from the competition by doing things that they value and build trust and credibility.

The right tactics for the right times

Here are some tactics that have helped us to get into the right sort of relationship with our prospects and clients.

E-marketing

The temptation is always to assume that the more information you send out, the more the recipient will be impressed. From the various campaigns that we have carried out, we have found that sending short, sharp bursts of relevant information can be very effective. But it is clear that it has to be newsworthy and easily digestible with a clear response mechanism if the reader wants more. The types of information that are always well-received are:

- Research and analysis.
- Market commentary and trends.
- Deals information.
- Team profiles.
- Plus the odd quirky or funny story, but not pictures of the internal staff party as that is not generally well received.

Our client research has indicated that you need to follow this up with a call from time to time to avoid the perception that this is indiscriminate spam.

Direct mail
For a slightly different approach, we tried this tactic. We picked a group of key prospects, and chose those where our contact information was up-to-date.

We sourced tins of good-quality biscuits. We sent them with a label suggesting a 10-minute catch-up over a cup of coffee. We delivered the biscuits and a few days later called the recipients to see if an appointment could be set up. By stressing that this would be just a 10-minute meeting, this helped us to get in front of the right people. Simple, effective and fun at the same time.

New media, including web films and podcasts
Sound bites and quick bursts of information delivered electronically are more likely to make an impact than many of the older forms of written communication.

We found that short web presentations can be powerful ways of getting our message across. In two minutes you can deliver a focused proposition in an innovative and involving way that is also entertaining. By sending clients and prospects a link to the website where these files could be accessed, we avoided many of the problems associated with fire walls and drive more traffic to our site in the process.

Simple ways to create a brand presence
Sometimes you just need to create some brand awareness. This can be as simple as taxis branded in your livery. For us, this exercise resulted in clients sending pictures taken on their mobiles, which were entered in a competition. Parking the taxis outside locations where corporate entertaining was taking place was also a good move. And hiring the taxis to take people to and from your meetings can create a good impression.

In summary, we have found at first hand that good targeting is vital. By and large it is better to work with a smaller group of contacts and ensure that you really get to know what they want from you and your team.

By taking a lively interest in what other organisations are doing, not just those in the property sector, you can develop an eye for marketing tactics that are newer and have a novelty value that can work to fulfil your business strategy.

Small firm case study: Chase Buchanan — tactical marketing

Michael Peacock is the manager of the St Margaret's branch and a partner of Chase Buchanan (*www.chasebuchanan.co.uk*), a residential sales and lettings agency established in 1994 by two partners that today has five offices in South West London (St Margaret's, Twickenham, Hampton, Kingston and London).

Image and advertising

When we started in 1994, we wanted our image to make us look established and professional. We updated it in 2004, as we had grown, and also wanted an image that was internet-friendly. We chose peacock blue and lime green as our colours and the new logo still resembles our original shield containing the intertwined CB letters, but with a far more contemporary look.

A little over three years ago, one marketing campaign, in our attempt to stand out, involved us putting "success" on our sale boards instead of "sold/sale agreed" and we linked the message into our advertising. We intended this to be a six-month campaign, but due to the response, we've continued with it ever since. Other agents have attempted to follow the lead with words such as "spoken for" and "sold it".

Community events

St Margaret's, just across the river from Richmond, is a relatively small, close-knit community and much of our tactical marketing activity focuses on strengthening those community links. Each year, we sponsor and take an active part in the annual St Margaret's Fair, as well as the summer and Christmas fairs at local infant and primary schools.

The St Margaret's Fair has been running for 30 years and this year Chase Buchanan were the main sponsors — receiving a full page ad on the inside front and back covers where we ran a competition to win up to £10,000 off our fees — with half of that going to charity. Entrants had to guess the combined value of 16 houses we had recently sold. For our advert in the programme we used a photo of a young boy jumping but with our corporate colours subtly in the background — it is also used in our brochure ("A short story about us"), which has no pictures of properties at all — it is young and fresh and focused on the sort of lifestyles our customers have or aspire to.

While some of the school events are cheaper, sponsoring The Fair costs almost £1,000. In addition, with staffing, refreshments and a number of "sponsored by" boards, the whole day isn't cheap, but is great PR. This year, a local café donated some cakes for our stall and we displayed their name in return. We email everyone on our database (several thousand clients) a few days before the event and encourage them to come and have a glass of wine or Pimms with us. It is primarily to have a chance to chat casually with people face to face — during the course of a transaction communications are usually telephone or email-based. It is rare that we see many of our vendors, unless we use opportunities such as this. We often pick up work too — valuations, lettings and sales — although we focus on promoting awareness and maintaining relationships at these events. We probably do not do as much canvassing or door-dropping as many agents.

Online promotion
The online space is increasingly important. We use Dezrez (*www.dezrez.co.uk*) — the main agent software, which allows our offices to be networked and uploads details of our properties to the main sites automatically and allows us to email and SMS those applicants who match any relevant properties. You pay per property so they're not cheap. The main site for our area is *www.findaproperty.co.uk* although we also use *www.homesandproperty.co.uk* (the London *Evening Standard* newspaper site) and *www.fish4homes.co.uk*. In Kingston, we have found that *www.rightmove.co.uk* has a reasonable response. Probably around 40% of those registering with us come through the portals, but a significant number still come from *Folio*, our own mini-mag, local papers and boards.

Memberships
We are members of the Guild of Professional Estate Agents (*www.propertyplatform.co.uk*), which has only one member in each area. They provide help with producing and distributing our monthly *Folio* newsletter and they place advertisements in the Sunday and London papers and put us in touch with the relevant journalists when they are producing articles on particular housing themes. We also belong to all the other main bodies, such as Association of Residential Lettings Agents, The Ombudsmen of Estate Agents and the National Association of Estate Agents.

Opening a new branch
We learned some hard lessons about marketing when we opened our Kingston branch. We wrongly assumed that our excellent reputation would carry itself across the river and that we could adopt the same approach as we had in other areas. However, we opened at the same time as a rival agent and although we had a High Street address, we were the wrong side of Kingston to really attract the family market. We walked a lot of streets, combined with a strong advertising campaign but, in retrospect, we should have done more homework on location and developed a "big bang" approach, such as 0% for the first month or two.

Importance of a strong service and relationship focus
While tactical marketing is important to us, so is strategic marketing — we place great emphasis on the service we provide. The people in this area are sophisticated and generally fairly wealthy and as such they want to deal with high-calibre people with experience. There are no weak links in our team — all of the selling staff in this branch have been managers themselves and therefore can identify opportunities quickly and provide a quality service to the client, who is

never fobbed off. They all have strong personalities and maintain the perfect balance between being friendly and professional.

Furthermore, we place a lot of emphasis on maintaining an ongoing relationship with all our applicants, buyers and vendors, which is the client service element of the mix — we make a particular effort to stay in touch with people throughout the 12 or 16-week contract period — many agents make a big play for customers at the outset but are less attentive once they have signed them up. We have used various completion gifts, such as bottles of wine and boxes of hand-made chocolates. We promote the fact that 97% of our customers would recommend us or use us again.

Selling and business development (sales strategies)

Roadmap

- Dealing with enquiries — from marketing to selling
- Targeting
- Conversion — the sales process, pipeline and cycle
- Decisions, decisions — the buyer's perspective
- Rainmaking — core selling skills
- Cold contact
- Networking nuggets
- First impressions — non-verbal communication
- Getting to know you — rapport and empathy
- Personality and communication styles
- Structuring a sales meeting
- Nobody expects the Spanish Inquisition — asking the right questions and active listening
- The psychology of persuasion
- Perfect presentations
- Following up
- It's in his pitch — pitching and tendering
- Negotiating — competition or collaboration?
- Closing the deal and handling objections
- The trusted adviser

"Seek first to understand and then to be understood."
Stephen R Covey

"Oh Great Spirit, grant that I may never find fault with my neighbour until I have walked the trail of life in his moccasins."
Cherokee Indians

"The deepest principle in human nature is the craving to be appreciated."
William James

These are some of my favourite quotes. Together they equip you to understand the basic principles of good professional selling. I am nervous as I embark on writing this particular chapter more than any other and this is for two reasons. First, while their surveying friends may need a little help in this area, most agents are supreme salespeople — whether it is because of their sheer force of personality, their years of experience, their incredible contact books, their love of the deal, their time spent learning from great seniors, their drive to meet ever-increasing targets or pure determination and persistence. Those who sell properties contain both the very best and the very worst salespeople. So it is hard to imagine what I can contribute — except that I did my first professional sales training almost 30 years ago and I have been training, learning, consulting and writing on the subject ever since. It is a passion.

The second reason is that the first book I wrote in 2000 was on the topic of selling skills for the professions. I was the first person to write an introductory sales book for lawyers, accountants and surveyors who have to tackle the professional and legal regulations that define their client relationships. Having written an entire book on the subject — and many hundreds of articles since then — how on earth do I extract just a few thousand words for this chapter?

My original training was as a psychologist. And this knowledge stood me in great stead for effective selling. Because selling is all about people. If you understand how people think, feel, perceive, communicate, form opinions and relationships and choose then you are half way to being an effective salesperson. I hope to share some of my psychological insights in this chapter.

Dealing with enquiries — from marketing to selling

If you remember when I introduced the subject of business development back in chapter 4, I explained that marketing was the various processes designed to raise your profile and generate enquiries. Selling takes over when you have to interact with the person or organisation that makes contact with you as a result with the purpose of creating trust and a relationship and converting that interest into business. Of course, once you have created a client the selling must continue — but it is slightly different, which is why I have placed all the material about working with existing clients in the next chapter on account management — which I call "clientology" — the science of working with clients.

If your marketing is effective you will prompt potential clients to get in touch with you. Too often firms don't realise how important this first encounter is. Clients sometimes ask about prices and surveyors and agents get defensive or talk solely about prices. But clients often tell me that they do not know how else to start a conversation or have better questions to ask. But clients are often not seeking the information they request — they are hoping that this short, initial interaction will help them decide whether or not they trust you and want to conduct business with you.

So look carefully at how you respond to different types of enquiries. Measure the success or conversion rate of different people in your office. Some may be better at converting telephone calls while others may be more effective at face to face. Watch who is most successful — and in some firms they have found that it is the support and secretarial staff rather than the professional staff — and see what you can learn about what they say and how they say it.

Rather than leaving this vital activity to chance, focus on it and see how you can quickly improve your sales effectiveness without too much time and trouble.

Targeting

Hopefully, as part of your strategic marketing plan (see chapter 4) and your tactical marketing campaigns (see chapter 5) you will have identified the particular segments of the market that you are trying to reach and the specific types of individuals and organisations that you want to talk to. The more specific you are in your targeting, the easier it is to make contact with those people having undertaken the necessary research to understand what they need and what messages might appeal to them and the best channels to use to reach them.

You can develop lists of triggers and filters to help you identify when your targets might be most amenable to an approach. Triggers are things that indicate an organisation is more likely to need external help — in the form of property advice in your situation. You should develop an understanding of the triggers that are most likely to lead to productive discussions in your target organisations. For example:

- Have fast-changing technology — changing their space requirements.
- Are affected by legislative change — leading to fast growth or market contraction.

- Face tough competitive pressures — meaning that they may have to enter new territories or reduce their overheads or are looking for ways to outsmart the competition.
- Have appointed new senior people — who may want to alter the way things are done.
- Plan significant growth — they need help with dynamic space planning.
- Have current advisers who are unlikely to be able to provide the depth or breadth of service that they need.

Similarly, it can help you narrow down your search and your targeting if you have some filters that help you sort out those organisations that have short-term potential and those that are going to spend much more time and effort before they identify a need with suitable motivation to do something about it. But typical filters might be (and you should develop your own list):

- Know people already.
- Can obtain a high level introduction.
- Have particular expertise that fits their needs.
- Have a good match in the size, location or nature of your firm.
- Understand already the decision-making process.
- Have experience in the sorts of issues that they face and have helped similar companies.
- Have relevant market sector experience.

The matrix in figure 6.1 may also help you bring some structure to the priority you assign to different targets.

Figure 6.1: Strategic decisions about which clients to target

	Low Desirability of client	High
HIGH Strategic Fit	Be pragmatic but sensible	Key contact TOP PRIORITY?
LOW	Don't even think about it!	Spend time checking it out

Source: Dr Cliff Ferguson (adapted)

Checklist and actions

- Do you have good systems for handling, converting and tracking enquiries?

- Are you confident that your marketing plan has properly identified your target market and ideal target clients?

- Do you have lists of triggers and filters to help you prioritise key targets?

- Do you have a prioritised list of target clients?

- Are your marketing activities generating the right amount and quality of sales leads?

- Have you set out some clear and SMART sales objectives for the firm, each team and each partner?

Conversion — the sales process, pipeline and cycle

Except in some residential or consumer situations, most sales cycles in the property industry will take time. You may need to contact, interact with and meet a variety of different people over a period of time before you are able to make the sale. The more money that is involved and the bigger and the more strategic the decision, and the more alternative suppliers that the organisation is considering, the longer the sales cycle is likely to be. Some research indicated that the commercial sales cycle is, on average, nine separate contacts over 18 months. You have to think about selling that way as a kind of courtship.

It might be worth spending some time thinking about the typical sales cycle for each of your different markets or services — as it is likely to vary depending on what you are selling and to whom. A key part of your sales management activity will be on working out how to reduce the time of the average sales cycle as well as improving your conversion rate (the number of actual clients or pieces of work that are generated from the leads).

It helps here to think about how your profile raising, general marketing and other activities support the flow of qualified prospects into your sales pipeline. Some call it the sales funnel. Some firms are very good at tracking their opportunity pipeline and can calculate, based on past conversion rates and a solid knowledge of their typical sales cycle, the likely future workflow or order book. This makes it much easier to do the relevant resources planning — having the right people in place — as well as reducing the sleepless nights caused by "What happens when the present big deals are completed" thoughts.

Figure 6.2: The traditional pipeline concept

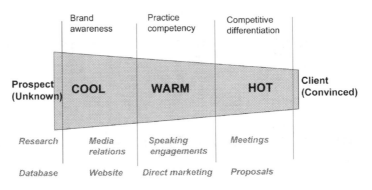

Integrated marketing communications

It is also helpful if you can set out the stages of the typical sales process for your clients and your services. Again, this will vary from firm to firm, market to market, service to service and client to client. But having a general understanding of the typical stages that you must go through will help you structure and plan your time and work, and also enable younger members of the team understand that Rome wasn't built in a day.

I spent some time developing a framework to try to help firms understand the different processes involved in managing a sales cycle and figure 6.3 outlines this but I always recommend that firms devote some time to thinking about the sales process that they are used to — and particularly comparing notes between different partners — so that you come up with a "best practice" approach.

Figure 6.3: Sales activities for a surveyor

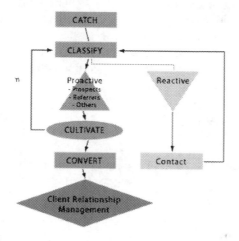

Source: Kim Tasso

Another popular model in the professions is that developed by the PACE Partnership, which is shown in figure 6.4.

Figure 6.4: PACE pipeline model

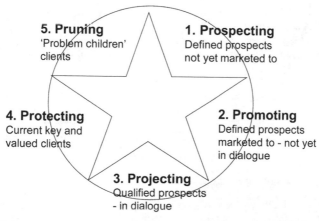

Source: PACE Partnership

Since the sales cycle can be long and protracted, it is probably a good idea for you to devise some sort of system to keep track of all your targets (including those among your key clients and referrers). I have seen highly effective systems in manual files, simple spreadsheets (such as the one below) and I have seen them in super-sophisticated CRM systems. The system you use is not important, the key thing is that you have some way to organise yourself that supports the way that you work and ensures that you maintain regular contact with your prospects and move them as efficiently and effectively as possible from prospect to client.

Table 6.1: Example client and prospect monitoring system

Name	Organisation	Sector	Status	Next action
J Brown	BigCo LLP	Professions	Client — Major	Lunch/dinner
J Black	PC Kicks	Technology	Client — Major	Send information
A Smith	Makealot plc	Manufacture	Client — Major	Introduce to partner X
P Halls	Beaming Ltd	Service	Client — Target	Invite to seminar
J Kingsley	Bigstuff Ltd	Manufacture	Client — Target	Speak referrers
A Jones	PartnersRUs	Professions	Referrer — Major	Refer work
H Brown	BigHitters LLP	Professions	Referrer — Target	Research
N King	Downtheroad LLP	Professions	Referrer — Major	Invite to event

Source: Kim Tasso

Checklist and actions

- Do you understand the different sales cycles within the various parts of your business?

- What is the typical length of the various sales cycles — how can you improve them?

- Do you have a system in place to measure what opportunities there are in your sales pipeline?

- Do you measure your conversion rate?

- Have you shared best practice with your partners on the most effective sales techniques?

- Do you have sales training programmes in place to help young surveyors and agents learn quickly and effectively?

- Does your firm have the right strategies, policies and processes in place to support effective selling?

- How do you manage, measure and monitor sales effectiveness in your firm?

Decisions, decisions — the buyer's perspective

While it is important to have a good understanding of the sales pipeline, process and cycle in the various parts of your market and for different types of client and service, it is also important to see things from the client or buyer's point of view. One of the main reasons that sales efforts fail is because the salesperson is too focused on their own need to sell something and neglect the needs of the buyer.

The first idea is the need for you to have good emotional intelligence — this is the ability to perceive, integrate, understand and reflectively manage one's own feelings and other people's feelings. There are five aspects to emotional intelligence:

- Know one's emotions (self-awareness).
- Manage one's emotions (handling feelings).
- Motivate oneself.
- Recognise emotions in others (empathy).
- Handle relationships.

The next thing is to understand the complex decision-making processes that clients go through when purchasing professional services. The following diagram highlights the various stages but, in reality, things do not progress in a nice, ordered way like this. These sorts of models do at least help you think about things in a structured way. For example, the best place for you to be is at the start of the process — helping the client identify and define the problem — so hopefully their solution specification focuses on those areas where you are most likely to help. The later in the process that you become involved, the more difficult it is to modify their agenda and present your solutions most favourably. When you are asked to pitch or tender, they are already at the evaluation of alternatives stage and it is much more difficult to differentiate your firm and beat the competition.

Figure 6.5: A client's decision-making process

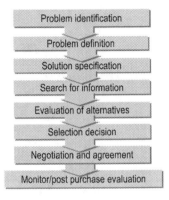

You should also recognise that there are various buying situations. If someone is embarking on a new task purchase (for example, looking at relocating their head office for the first time) they will need a different sort of sales strategy to if they are undertaking a modified rebuy (for example, looking at help for a lease negotiation when they have already been through the process once or twice before) or when they are in a routine rebuy (for example, a retailer leasing and fitting out its 50th store).

Something that some professionals find hard to appreciate is that many purchasing decisions are not made on a purely rational and logical basis — who has the best team, expertise, offer and price — as you would expect. Because people are emotional and irrational creatures, a whole host of stuff becomes involved in the decision process — even if they believe and tell you that they are making a rational decision. Some of these emotional factors will relate to how comfortable they feel with the people in your team, their level of risk aversion, how reluctant they are to change current processes and procedures, how secure they feel in their roles, how they think their peers will perceive their choice, internal politics, your and their reputation (or brand), their motivation and also cognitive dissonance (in effect, if you decide to appoint new advisers it means that your present advisers are not so good — and if you have chosen them then you are more likely to think that they were a good choice in the first place!).

Another valuable model is that of the decision-making unit (see figure 6.6). Particularly in commercial property situations, you are rarely dealing with just one individual at an organisation. You may have to deal with site managers (users of the service), the finance director (one of the decision makers), the in-house lawyers and purchasing professionals (the buyers) as well as the real estate director on the board (the decision maker) and his assistants (gatekeepers). They may also be keen to take account of the views of influencers such as their accountants, management consultants, present landlords or bankers and there may be people working in their organisation for you (sponsors) and against you (anti-sponsors).

Figure 6.6: The decision-making unit

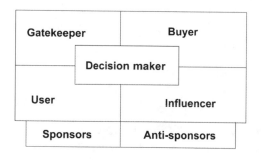

Source: Webster & Wind

Mapping out all the people that you know (and need to know) at an organisation and developing a strategy to deal with their different needs in terms of information (some will be real estate experts like you and others will be relatively inexperienced in the subject) and the nature of the relationship they need is a great way to build a sales strategy for an organisation (see page 282).

Rainmaking — core selling skills

Perhaps it is my Native American routes, but I like to think of great salespeople as rainmakers. Native Americans are skilled hunters and spend much time researching to understand their prey. But they also have great respect for their prey — only targeting those that they need, and honouring their kills and ensuring that every piece of any animal is put to good use.

So what are the core selling skills that you need to be an effective rainmaker? There are numerous skills — many of which are touched on in this chapter. But a particularly elegant way of considering them was developed by my friend Dr Clifford Ferguson as a result of many years' research and field sales experience.

Figure 6.7: Rainmaker assessment

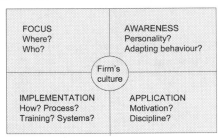

Source: Dr Clifford Ferguson

Figure 6.7 indicates that the first requirement is for rainmakers to have sufficient focus to know whom they are targeting and how to reach these people. Having a good marketing strategy should support them here. Then they need to understand their own personality — its strengths and weaknesses — and how they must adapt their behaviour with different types of clients in different situations. There is some material to help with this in chapter 8 on people. They also need to understand the sales process and what behaviours, systems and activities will be effective — this is part of their sales training. Finally, they must be motivated to apply themselves to developing new business and have the discipline to dedicate the necessary time on a regular basis to doing so.

These are general guidelines only. I have invested time in many property firms working with the best "rainmakers" to develop competency frameworks of the attitudes and behaviours that can then be used as a foundation for sales training programmes.

Checklist and actions

- How is your emotional intelligence?

- How well do you understand the different decision-making processes among your various client groups?

- Do you modify your sales approach for the different types of buying situations and motivations of your target clients?

- Do you understand the roles of the different people in the decision-making unit at key targets?

- Do you have written sales strategies for how you will develop relationships with the various players at your key targets?

- What are the key rainmaking or selling skills and competencies at your firm?

- Are there suitable training programmes to help your people acquire and practice the relevant sales skills?

Cold contact

The concept of cold contact generates a lot of heated debate among professionals. There are some that think nothing of picking up a phone or sending an email or letter cold to someone to try to win an appointment. There are others who advocate using the gamut of marketing communications activities to first raise awareness, educate them about your firm and its services and generate low-commitment opportunities to meet with them — for example, at an exhibition, conference or seminar — before trying to persuade them to commit time to a meeting.

If you are writing a letter first, use one of the frameworks like AIDA (see chapter 5 on tactical marketing) although often a detailed proposal (that you have researched so that you demonstrate your knowledge of their organisation and the possible ideas and solutions you have to offer) will often be more effective for large organisations.

The key to effective telephone selling is preparation. The other is confidence — and a robust attitude towards rejection!

1. Advance preparation
   a. Research — the target organisation and the people you wish to speak to.
   b. Set objectives — for what you want to achieve from the call eg, learning a particular piece of information, permission to send a document, a commitment to a meeting).
   c. Identify a topic of interest — your research should reveal this and it should fit in with the service that you are promoting.
   d. Make a note of the key points and questions you plan to use — I don't advocate having a detailed script — it rarely sounds natural. But a sheet of key points to prompt you will be helpful and will boost your confidence.
   e. Rehearse — talking through — out loud — what you are going to say, your introduction, how you will ask and answer questions etc — will help you feel more confident.

2. Making the call
   a. Introductions — how you introduce yourself is important. A smile works down the telephone line. Some people stand up to make them sound more attentive.
   b. Gatekeepers — do your best to develop a good relationship with any personal assistants or secretaries who are screening calls. They can be an invaluable source of information and a great ally if you treat them with respect and they can make your life impossible if you upset them.
   c. Verify the speaker — check that the person you are speaking to is the correct person with the necessary brief and authority to deal with the matter about which you are calling.
   d. The timing of the call — check to see that you have not called when they are busy or in a meeting, and agree a better time to call back if they are.
   e. Generate interest — back to the various tools, but often a question will help or a reference to something about their organisation that you have noticed from your research.

f. Listen — carefully listen to what they are saying — they will provide vital clues on how to make the most of the opportunity and how best to proceed.

g. End the call — when you have achieved your objective (obtained the information you need, agreed a meeting date and time, agreed the way forward, promised a written document, determined a better time to call back etc) check whether they need any other information, thank them and close the call. Stay in control and avoid them ending the call.

3. After the call
   a. Make notes — Whether you are using a manual/paper system or a fabulous CRM system, ensure that you note the key points of the call and the agreed follow up action and enter any reminders or prompts into your calendar.
   b. Take action — Do what you say you will do. A short confirmatory email is also always a good idea too — it's professional and ensures that there are no misunderstandings. Clients like to know that their professional advisers are good at documenting all communications.

Handling telephone enquiries

Of course, it may not be you making that first telephone call. If your marketing works effectively you should find that potential clients are calling your office in response. Make sure that whoever operates your switchboard is aware of any campaigns and knows who to transfer calls to.

Much of the information in the section above will be relevant, but it will also help to have a checklist of the things that you must cover during such a preliminary call. As I suggested, speak to those in your firm who are particularly good at converting telephone enquiries into meetings and instructions — and see what tips you can pick up. Another point worth mentioning is often enquirers do not know what questions to ask — so they will ask about price — you should always answer such questions but spend some time at the outset developing some rapport, obtaining information and talking a little bit about them as well as yourself, the firm and the services you provide before you do so.

Figure 6.8: Possible structure of a telephone enquiry

Source: Kim Tasso

The advantage of using a diagram rather than a checklist to support calls like this is so that it is easier to imagine how time elapses and to speak accordingly. For example, imagine going through all these stages in a five-minute call compared with a 20-minute call.

Networking nuggets

So much business in the property sector is done through networking that it is both a vitally important concept and one that many will be very familiar with. So this section is really aimed those who are new to the topic — and to remind the seasoned, experienced networkers among you that younger members of the team might need some help to become as effective as they are.

Are networkers born or made? Obviously, having a particular type of personality will make it easier for some than others. If someone truly hates the idea of working their way through a room of hundreds of strangers, then find an alternative way for them to generate contacts and business. But great networkers blend the hard work of focused targeting and sustained follow up with the relaxed serendipity of interesting chance encounters.

Here is some guidance — but you need to have a strategic marketing plan that outlines who you are targeting and why and what you are doing to raise your firm's profile among the relevant groups before you set off on your networking programme.

1. **Before the event**
   a. **Manage your expectations**
      It is rare for results to be instant. You have to network regularly and consistently before you become adept at meeting and cultivating the right people and it takes a while to develop relationships with those contacts you make in a networking environment to become prospects and clients (see the material above on the sales cycle).
   b. **Do your homework/research**
      You need to prepare yourself before you go. Hopefully, this will be done as part of your strategic and tactical marketing.
      i. Read the business and marketing plan.
      ii. Identify your target audience and build a list of targets.
      iii. Select the type of event that will reach the right type and level of contacts you are targeting.
      iv. Know why are you going — be clear about your aims and expectations.
      v. Set objectives — set realistic goals for example, "I want to meet five people who work in XYZ type of business".
      vi. Know who will be there — get in touch with the organisers to see if they can tell you a bit about who attends, although due to Data Protection Rules these days they are unlikely to be able to send you a list of past and future attendees.
   c. **Practice your introduction**
      i. See the section below.
   d. **Think about things to talk about**
      i. Part of your preparation should be reading the national, property and local press so that you are familiar with current events and the things that people are talking about. You should also be familiar with the latest developments at your firm and among the target audience.
   e. **Adopt a PMA (Positive Mental Attitude)**
      If you feel confident and happy, this will be reflected in your non-verbal communication and it makes it easier for people to talk to you.
   f. **Develop systems to track progress and stay in touch**
      This relates back to the earlier material in tactical marketing on databases and CRM tools and the earlier material in this section.

2. **At the event**
   a. Approach people — look for "loose threes" or those whose posture, facial expressions and position indicate that they are keen for others to join them.
   b. Have a good introduction — be confident and positive.
   c. Make people feel at ease.
   d. Show interest/ask questions.
   e. Listen intently.
   f. Know your stuff.
   g. Gather information.
   h. Find reason to contact them again — while requesting a business card is fine, people may be suspicious unless you have indicated what you are going to do with it — send them some information, contact them for a coffee, put them on a mailing list, pass it on to a colleague etc. Remember that people are unlikely to agree to a high-commitment follow-up (eg, lunch, a meeting etc) after a brief networking meeting, so have some low-commitment ideas (eg, copies of articles, details of a contact that they would find useful, a regular newsletter addressing their area of interest or market, invitations to a social event) ready.

3. **After the event**
   a. Assess the value of the event — were the types of people there that you wanted to meet? Is it worth returning? How can you prepare better next time? Was there enough time to talk or do you need to get there earlier/stay later next time?
   b. Write notes — do this as soon as possible. Some networkers make notes on the back of business cards they have collected. Others use personal organisers, phones or the contact system on their laptop.
   c. Follow up immediately — within two days ideally but certainly no more than a week.
   d. Ensure you deliver on any promises.
   e. Contact again regularly (but don't bombard) — this is part of your contact management system.
   f. Build the relationship.

But what if you get "stuck" with someone? This is one of the most common questions I am asked. Obviously, you must ensure that you remain polite and do not create a poor impression, so here are some ideas:

- **Escape — What to say if no common ground ...**
  - "Come and meet my colleague ..."
  - "Shall we go over to ...?"
  - "Can I get you a drink?"
  - "Anyone particular you wanted to meet?"
  - "Really nice to meet you ... must circulate."
  - "Lovely to meet you ..."
  - "I'll send you the information."

- **Extract — What to say if reason to meet again is identified**
  - "I will send that information ..."
  - "I look forward to continuing this conversation ..."
  - "I will speak to you over the next few days."
  - "It's been great to meet you, I'm so glad we'll meet again."
  - "I will call to confirm ..."

- **Exit — What to do ...**
  - Use transitional words ("anyway")/break eye contact (down).
  - Join another group (and take them with you).
  - Get your colleague to rescue you.
  - But always leave in the nicest possible way.

Another tip is to remember "giver's gain". This is where networkers are recognised for always having something to give other people — putting them in touch with relevant people, offering support in a non-business context, always being prepared to help etc — even if there is no immediate benefit to themselves. This requires you to make a big investment in time to your contacts and also to take a long-term view of the benefits of networking.

Checklist and actions

- What is your policy about cold calling?

- Are your cold calls — whether by telephone, email or letter — as effective as they could be?

- Are there systems in place to manage cold-calling programmes, to record their results and to ensure proper follow up?

> - Have you reviewed the procedures and training for how inward telephone and email enquiries are managed?
>
> - Do you have systems to measure the conversion rates of different types of enquiries?
>
> - How well do your people prepare for networking events?
>
> - Are there systems in place for people to record the contacts made at networking events and to help them follow up in a timely manner?

First impressions — non-verbal communication

You only have one chance to make a good first impression is a familiar saying. And it is true. Psychologists have shown that people form an impression of you within a few fractions of a second of first meeting you — and the evidence suggests that those initial impressions rarely change. So there is something other than what you say when you meet people on which they base their impression of you.

Professor Albert Mehrabian, a communications researcher, found that the meaning of communication is transferred in the following way:

- 7% — What you say — the WORDS
  - Key words
  - Style of language (see NLP)
  - Common experiences and associations

- 38% — How your voice says it — the MUSIC
  - Accent
  - Tonality of voice
  - Pitch and projection (volume)
  - Pace and pauses

- 55% — What your body does — the DANCE
  - Your physiology and breathing
  - Posture and gestures
  - Facial expressions and, most importantly, smiles and eye contact
  - Handshake (neither too firm nor too weak)

Non-verbal communication is very important when we wish to develop rapport with people. It is also vital that you convey your confidence by adopting the appropriate posture, gestures and stance because even if you are not feeling very confident, you can learn to appear as if you are. This is covered in greater detail in the subject of Neuro Linguistic Programming (NLP), which examines the impact of both what you say and how you say it in developing good relationships and in persuasion.

Another way to think about first impressions is to consider your personal power — and its source. There are advantages and dis-advantages to each type, so you might think about developing other aspects to those where you are strongest.

Figure 6.9: Personal power

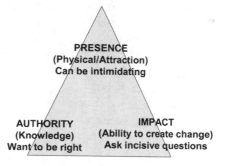

Essentially, your effectiveness at networking will depend on whether or not people like you. So the question has to be "What makes people like you?". Well, the more things you have in common with someone, the more likely you are to have some common ground and get on. But research indicates the following attributes:

- sincerity;
- physical attractiveness;
- similarity;
- contact;
- co-operation; and
- association.

A final point about networking is that it is good to hunt in a pack. Having a colleague come along with you makes it less scary and provides you with an opportunity to just chat now and again — rather than network. If you choose a colleague with a slightly different personality, style and area of work, then you are increasing the chances of finding something in common with those you meet and it is always easier to join other people when you have a mate with you. What's more, you might need them for some of the escape strategies mentioned above.

Introducing ...

The way in which you introduce yourself is crucial. Not only because you are conveying information, but also because you are providing hooks on which a conversation can continue. If you say "Hello my name is John Smith from ABC firm" there is little to go forward. However, if you say, "Hi, I am Kim and I work in management in the property industry and I live in Twickenham" the other person can ask about management, the property industry and Twickenham.

As well as having a good introduction, you might also think about preparing an elevator speech. This is a short description of your firm and its services, lasting no more than about 90 seconds, which you would give to a senior decision maker if you were lucky enough to get stuck with them in a lift for a short journey.

Getting to know you — rapport and empathy

Some people do not like selling. They feel it is somehow unprofessional and/or they are afraid of coming across as being too pushy. Often, the idea that selling is unprofessional is because we are usually only aware of selling when it is bad — such as when double glazing or timeshare salespeople launch into some long pitch without caring whether or not we are remotely interested.

Before we worry too much about what we say to people, we need to develop a deep empathy of their perspective — their needs and interests. Most professionals are very good at doing this, but then fail to initiate the dialogue about how they can help people. Sometimes, it is easier to think of selling as "helping people to buy" or "solving their

Figure 6.10: Why don't we like selling?

	EMPATHY	
PROJECTION	**Low**	**High**
High	High pressure selling	Good relationship - with sales
Low	No dialogue	Good relationship - without sales

problems" and this is the idea that underlies most of the material in this chapter — it is called "consultative selling", as opposed to the old fashioned way of more forceful projection.

Empathy is a key skill to help you get on with people. And non-verbal communication (see above) can help you develop a rapport. Rapport is one of the most important features or characteristics of unconscious human interaction. It is a commonality of perspective, being in "sync", being on the same "wavelength" as the person with whom you are talking. There are various techniques that are supposed to help in building rapport, such as matching or mirroring your body language, maintaining eye contact and matching breathing rhythm. Many of these techniques are explored in Neuro Linguistic Programming. Research indicates that most people achieve rapport naturally with between 1% and 30% of the population and with training and development most people can double their starting percentage.

Personality and communication styles

Knowing the nature of your own personality and how it impacts on your ability to get on with people with different personalities is vitally important in selling and client management. A number of frameworks and models are discussed in chapter 8 on people.

You should also be aware that if you are used to being a leader or an expert, you may naturally adopt a push approach to communication, whereas when you are meeting people for the first time or trying to develop trust and rapport then a pull style of communication is often more effective. This is covered further in chapter 7 on clientology.

Checklist and actions

- What sort of first impression do you, your partners and your staff make?

- How well does everyone understand the importance of non-verbal communication both in new and existing client situations?

- How do you introduce yourself, your colleagues and your firm? How good is your elevator speech?

- How efficient are you at generating empathy and rapport with all types of people that you meet? How might you enhance your skills with those who are less like you?

- Have you read the other parts of this book concerning the importance of personality and communication style?

It's all about needs

The earlier sales process model focuses on the critical idea of selling being concerned with identifying and working with people's needs. Without a need, there can be no sale. So much of the initial work in developing a client relationship is asking questions and learning about your prospect's needs.

But it is too easy to focus on their immediate property-related needs — which is where you want to sell. So be aware that you need to explore a wide range of other needs before you see the full picture and can hone in on the most important and immediate needs where there might be an opportunity for you. I find it helps if you use a table to try and find out and explore the various needs of each person in the decision-making unit.

Table 6.2: Mapping out the different needs of your prospect

Personal needs	Role needs	Team needs	Organisational needs	Industry needs	Other needs

You use questioning techniques and good listening skills to obtain this information.

Structuring a sales meeting

Let us now assume that your marketing, your cold letter or telephone approach or your networking has resulted in to your prospect agreeing to a meeting. Also let us assume that you have about an hour at that meeting. How are you going to prepare for that meeting? And how do you plan to use the time in that meeting?

Most good salespeople will tell you that their success is mostly down to hard work — doing the research, preparing well, thinking things through in advance, talking to their colleagues for ideas and so on. So a couple of days before the meeting allocate some time to do the necessary research and preparation, and decide how you plan to spend your time there. Obviously, the time plan will be different if it is the first meeting with a brand new contact to the fifth meeting with a client where you are trying to finalise a deal.

Also remember that you need to drop them a line to confirm the key points agreed at the meeting. You might try a "4Cs letter", which confirms the points discussed, clarifies your understanding of the situation and needs, explores the collaboration potential and confirms the next steps. If you are at the appropriate stage of the sales process, this could also be an outline work plan showing what, who and why is to be done and by when at what cost and reminds them of the desired outcome, results and benefits. And the reason a follow up note is so important? It shows respect for their time and that you are responsive. It avoids misunderstandings and acts as a valuable aide memoire for them (which they may use to communicate with others in their organisation). Many clients have told me that these sorts of letters help them structure their own thinking — so they value them a great deal.

Nobody expects the Spanish Inquisition — asking the right questions and active listening

Chinese proverb: He who asks a question is a fool for five minutes;
He who does not ask a question remains a fool forever.

As I have mentioned above, the key to effective selling is to have a good empathy with your prospect and to fully explore their current and potential needs. To do this, you must ask questions. We were all taught about the difference between open questions (Who? What? Where? Why? And How?) and closed questions (ones that can be answered with Yes or No) and their important role at different stages of selling.

But it helps to think about the different types of questions that you might ask, the order in which you use them during a meeting, and even developing a list of possible questions in advance.

- **Broad questions to identify key issues**
    - Helpful in promoting dialogue and scoping.
    - Reveals all possible issues that can be explored in greater depth before more specific questions to understand in detail.

- **Open questions (Who, Why, What, Where, How) to elicit information**

- **Closed questions (do, will, shall ...) to test commitment**
    - To gain decisions or agreement.
    - To reveal objections.

- **Descriptive scenario questions to demonstrate knowledge of the sector**
    - By describing a situation and asking how it relates to them.
    - This also reassures them and can add value.

- **Combined questions**
  - Where you ask two common questions together to force them to think about an issue from a different perspective (eg, "No doubt you are experiencing the same problems caused by the credit crunch as everyone else in the retail sector, but how are they particularly affecting your business and what solutions have you been looking at in the areas of technology and distribution?").

- **Summarising questions — to show that you have listened and understood**
  - To give you time to think and reflect.
  - Provide an opportunity for them to clarify/confirm.

I advise people to use questions to promote divergent thinking (looking at a topic in the broadest possible sense) before summarising the key issues with more convergent thinking (bringing together the various strands). This is a valuable tool in creative thinking (there is more on this topic in chapter 3 on leading).

Many sales training programmes focus on the different types of questions that you ask and help people prepare and practice using those questions to increase their sales effectiveness. Copyright issues prevent me describing many of these models, but the following shows my framework and some of the most commonly used ones.

Figure 6.11: Seeing STARS

Situation or facts
Tackle the underlying challenge or problem
Assess (relative) importance
Realise what benefits/values in a solution
Suggest the next steps

Here are some other approaches:
Situation **P**roblem **I**mplication **N**eed/payoff (Huthwaite International)
Surface **H**unt **A**djust **P**aint **E**ngage (Divdale and Lambert)
Opportunity **R**esources **D**ecision **E**xact solution **R**elationship (Khalsa)

Source: Kim Tasso

And my comment about the Spanish Inquisition? Well, you should not get so carried away with asking all the questions that you forget about sharing some information about yourself to balance the exchange. Otherwise, the other person will feel as if they are being "processed".

Silence is golden — active listening

The obvious part of asking good questions is to make sure that you listen properly — ideally, for at least 50% of the time — to what the other person is saying. Active listening is a skill that we must practise. Too often, people just stick to their own agenda and fail to pick up important clues and comments from the other person so that they can modify what they say to improve the exchange and boost their effectiveness. Here are some pointers for active listening:

1. Sit or stand with an open stance (no folded arms or barriers).
2. Maintain eye contact.
3. Summarise key points on a regular basis.
4. Provide continuation prompts (eg, "Yes", "Then what", "I see" or even a smile or a nod).
5. Ask questions about points that have been made, which shows that you are listening and interested.
6. Take notes — but only if this is possible.
7. Ask for clarification/more detail.
8. Avoid preconceptions — your mind is very effective at ignoring or filtering out information that does not fit with your inner model or concept of what the other person is like or concerned about — so keep an open mind.
9. Think about what makes a good listener — there will be people around you who are either great or poor listeners, model your behaviour on the good ones.
10. Make sure that you can hear — sometimes you may need to move away from a source of noise, or ask someone to speak up if you are having difficulty paying attention.
11. Use your eyes — eye contact is used to cue when people should speak.
12. Do not "switch off" — if you feel you are losing concentration, then ask a question or summarise a key point.

If you ask a question, do not be tempted to start talking if they do not respond immediately. Sometimes people need time to think and gather their thoughts. Be comfortable in the odd moment of silence.

The psychology of persuasion

How do we persuade people to do things? There is an old saying that goes something like "You say it — they doubt it. You argue it — they defend it. You prove it — they diminish it. They say it — they believe it". So the answer is simple, you just need to get them to tell you what it is that you want them to do — as if it were there idea.

There is much material in psychology on this topic. For example, is it better to present just one side or both sides of an argument? An experiment found that a two-sided presentation was more effective with educated people (who are trained to resist a one-sided argument) but a one-sided presentation is more effective with less-educated people. Also, we need to remember the primacy and recency effects — the fact that people are more likely to recall those items presented first or last (this is particularly important in presentations and pitches). If you are using a single source, then the primacy effect takes most importance in a discussion whereas with multiple sources it is the recency effect.

Another useful tip is that, in order to convince other people, you need to convince yourself first. This suggests that when you are trying to win over one of your partners, you ask him or her to persuade someone else and the job is practically done.

Most ideas about persuasion presuppose that you know how someone's mind is made up at the outset — otherwise how do you know in which direction to change them. In essence, you help them move towards your position by creating doubt and thus generating movement. A key tool here is the use of stimulating questions, to help people evaluate, speculate and innovate. In the section above on questions we saw how they can be used to explore needs, consider the benefits of proposed change, test reaction to proposed terms, consider the value and benefits of change and even build support.

I find the MUSCLE acronym helpful to remember the various ways to persuade people:

Mutually advantageous concessions
Understanding genuine needs
Strength: Power and Coercion

Compromise
Logical argument
Emotion

There is a great book written by a fantastic psychologist that is quite readable and accessible to non-experts called *Influence: The psychology of persuasion* by Robert B Cialdini. He points out that much human behaviour is based on fixed action patterns, which are effectively automatic responses generated by a single trigger feature. He gives plenty of examples for those who doubt him, for example, when requesting a favour you will be more successful if you provide a reason — "because" is a trigger.

Checklist and actions

- Do you devote adequate time to planning what you want to achieve and what you will do at each sales meeting?

- Do you always prepare an agenda for each sales meeting?

- Do you always send a confirmatory note — setting out the next steps — after a sales meeting?

- Have you worked through a list of the different types of questions that you use in different sales situations?

- Have you identified those questions that generate the most information and the best impression?

- How good are you at listening to what people are saying — and modifying your approach in response accordingly?

- How do you typically try to persuade people — and what other techniques and approaches might you attempt in future?

USP or features and benefits

There is much in the old fashioned sales training books about the need for a USP (unique selling proposition). However, the idea that you can find one single outstanding aspect of your property service that is unique and different to everyone else is rather a stretch. When it comes to presenting your case in a sales situation, it might be easier to think of developing a series of features and benefits that together make the ideal solution for your client. In order to do this, we need to understand the difference between a feature, an advantage and a benefit:

Features	Advantages	Benefits
We have worked for over 100 clients in ...	• Lots of experience • Knowledge of best practice • Shorter learning curve • Less chance of errors	• **You** obtain a quicker or cheaper service • **You** get policy input or ideas • **You** feel secure
We are one of the largest firms in ...	• Wide range of skills • Lots of resources • Economies of scale	• Easier/more convenient for **you** to have a "one-stop shop" service • Always someone available for **you** • **Your** work gets done faster

You will notice that usually features statements are inward looking and start with a "We". You get to the advantages and benefits to "You" the client by challenging them with the question "So what does this mean to me?". This is a good exercise to apply to your brochures, websites and proposal documents.

A related idea in selling is that of a value proposition — which should answer the questions: "Why should I buy this product or service?" as well as "Why should I do anything at all". Value, as with beauty, is often "in the eye of the beholder" (eg, we each place different importance or weight on things). In essence, a value proposition is an offer to some target in which they gain more than they give up (in terms of merit or utility), as perceived by them and in relationship to alternatives, including doing nothing (opportunity cost). A value proposition is generally a clear and succinct statement (eg, two to four sentences) that outlines a firm's unique features to potential clients.

Perfect presentations

I could write an entire book on preparing and delivering perfect presentations but I have to select just a few key pointers for this small section of this one chapter. It is a good point to make though because often we try to cram so much stuff into our presentations that we lose our audience along the way by hitting them with too much information and failing to stress the key messages. For great presentations, you need to remember "less is more".

In my presentation training sessions, I use the following framework to address all the critical issues in developing a perfect presentation:

- Planning
  - Consideration
    - The purpose and aims of the presentation.
    - Different types, formats and styles of presentation.
    - The marketing and selling basics.
  - Content
    - Structure of the presentation.
    - Different types of audience and involvement.
    - Handouts.
    - Procedures (to establish in your firm to provide support for those producing and following up presentations).

- Performance
  - Confidence
    - Non-verbal communication.
    - Managing nerves.
    - Communication aids.
    - Using the physical environment.
  - Conversion
    - Sales process.
    - Networking.
    - Following up.
    - Developing relationships.
  - Consolidation

Let us start by considering the structure of your presentation, there are numerous options:

- Problem/solution. Many property people use case studies and do before and after to get their point across.
- Chronological. This is often used to talk about major developments, such as regeneration projects.
- Spatial. This depends on how the topics link together (see below).
- Topical. Linking to present market conditions or recent cases or issues.
- Theory/practice. A common approach in property — looking at legislation and regulations and how you apply them in reality.

The Aristotle approach is more akin to storytelling:

- Establish the scene to draw in the audience.
- Make the audience a character in the story.
- Describe the conflict or issue to be resolved.
- Explain the recommended goal or pay-off.
- Propose a solution.

Then there is the old favourite "Tell 'em what you are going to tell them. Tell 'em. Tell 'em what you have told them". As well as being good signposting, this also has the effect of reinforcing the key messages.

When preparing your talk you must consider the time available. Most people find it harder to develop a good presentation lasting just 10 minutes than they do one for an hour. A rule of thumb is no more than one key point for every five minutes and a maximum of five points.

Often, I will use a mindmap to try and consider all the various points I wish to make — and then look at the links between them. This approach has the advantage of breaking you out of linear thinking and providing a way to group and categorise key elements. There is some great software (eg, Freemind) that enables you to produce mindmaps easily on your PC.

So we have a structure and have organised the content. Now we need a really impactful start to things (remember the point above about first impressions being important?). Here is a selection of ideas to make things start with a bang:

Attention — Capture their attention.
Benefits — What will they gain from listening?
Credentials — Provide your credentials (this is particularly valuable for international audiences).
Direction and Destination — Explain your structure.
Empathise, enthuse, energise and enthral!

If you are providing a longer presentation, be aware that people can only concentrate for a short while.

Figure 6.12: Concentration curve of audiences

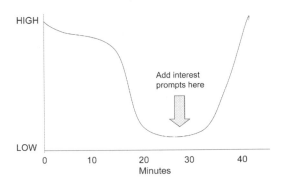

Therefore, you need to add interest — through the use of interesting visuals, by asking the audience questions, by getting them to do something, by using a different type of media or changing the presentation style or pace. Ideally, this should be done after each 20-minute chunk in order to keep the audience alert. You should remember that our brain works at 500 words a minute yet we listen at 150 words a minute so if we are not careful our audience will switch off.

I have seen many presentations that are just words and bullet point lists that seem to be more to support the speaker than they add value or interest to the audience. If you are going to use visual aids then make them interesting. The nerves from the eye to the brain are 25 times larger than those from the ear to the brain and 85% of information is absorbed visually. A clear case of a picture being worth a 1,000 words. With the facilities in Google Image and Microsoft and the ease with which you can take high-quality digital images, there is no excuse not to use attractive images in your presentations. Also, bear in mind that human perception is selective. People will automatically focus on things in the following order: images of people's eyes, images of people, images of objects, headings, signatures, postscripts, text containing recipient's name/familiar number and short positive words.

Here is a short list of some visual aids and alternatives to Microsoft Powerpoint.

- Hard copy of PowerPoint slides.
- Other hard copy materials (bound):
  - charts;
  - tables; and
  - flowcharts.
- An expanded agenda.
- Pre-prepared or spontaneous flipcharts.
- Printed boards.
- Websites and extranet demonstrations.
- Videos.
- Webcams.
- Teleconferencing.
- Role plays and sketches.
- Debates.

One of my pet hates is poor timing. I become cross when speakers over-run their slots. It shows poor preparation, a failure of self-discipline, an absence of rehearsal, an inflated sense of importance and a lack of respect for the audience and the poor chairperson who has to try and maintain the overall schedule. If you have a 30-minute slot, prepare enough material to cover 20 minutes, and then you are likely to finish just before your 30 minutes are up.

Public speaking is one of life's great terrors for many people. You must recognise that while you might find addressing a group of people a completely natural thing to do, it really does terrify some people. If you suffer from stress and nerves, seek some help. When you are afraid your body releases adrenaline and cortisol into your blood stream, which have a real physical impact on your breathing, temperature and your voice (often you will find your throat dries up). Visualisation techniques help some people:

- Prepare
  - Find somewhere quiet.
  - Calm yourself and clear your mind.
  - Breathe slowly and deeply.

- Focus on your goal/desired state
  - Visualise yourself having achieved the goal/state.

- How do you know that you have achieved your goal/desired state?
- What have you done or achieved?
- What does it look like? Colours? Shapes? Situations?
- What does it sound like? Voices? Words? Sounds?
- What do you say? What do others say?
- What does it feel like? Emotions? Physical effects?
- What are the most pleasant aspects?
- What is different? What is happening?
- Spend at least five minutes exploring the vision.

- Capture the state
  - What are the most significant elements of your vision?
  - Fix key positive points and the most powerful images/sounds/feelings in your mind.
  - What would you do to recreate this state?

A key element in many presentations is to develop some interaction with the audience. Getting them involved by asking them to do things, requesting for a show of hands to obtain a feel for the mix of interests or even promoting debate among them will all help. In smaller group situations, where you might be pitching for business, you really want the audience to be as active a participant as yourself — so think about ways to promote dialogue and interaction in advance. I also suggest people "lighthouse" across the audience so that eye contact feels as if it is maintained with every audience member. This should also enable you to pick up tell-tale body language signs that all is well (or not), thus enabling you to change tack or prompt some action to fit the mood appropriately.

Often, you will be thinking about how you promote action after your presentation. Therefore, you will produce some form of handout for the audience. You need to consider whether you want people to use these handouts before, during or after the presentation. Naturally, you will ensure that the materials are well branded and consistent with your presentation and that your name and contact details are on all sheets. You may want to include additional reference material (although I always suggest that you encourage people to contact you for these as an aid to follow up), space for them to make notes, speaker biographies and photos, case study examples or even relevant marketing materials. Hopefully, you will also be brave enough to include a feedback form so that you can learn how to improve in the future and the particular areas of interest or relevance for them.

Here is an overview of the full range of topics that are relevant. Many of them are addressed elsewhere in this book. Happy presenting!

Presentation checklist

1. Consideration
   a. What are your aims? What do you want to achieve?
   b. How does this presentation fit within your overall marketing/sales plan?
   c. Who is your target audience?
   d. How will you develop empathy with the audience?
   e. What does your target audience want to get out of the presentation?
   f. What lasting impression do you want to create?
   g. What advance preparation do you need to do?

2. Content
   a. What are your key messages? (and how does this link to your business aims?)
   b. How will you structure the information? How will you signpost progress?
   c. In what order will you present the information?
   d. Will you use visual aids (such as Powerpoint or flipcharts)?
   e. What technology will you need? (Projectors? Microphones?)
   f. What facts and figures or images do you require?
   g. What can you omit if you run out of time?
   h. How will you achieve impact at the start and end on a high?
   i. Does the amount of material fit the time available?
   j. Will there be time for questions?
   k. What sort of handouts will you provide? Before or after the presentation?

3. Confidence
   a. Have you rehearsed and practised your presentation?
   b. Have you checked your appearance?
   c. Are you familiar with the room and equipment where you will present?
   d. Have you allowed yourself enough time to arrive calmly and prepare?
   e. What sort of first impression do you want to create?
   f. Will you use notes to prompt you?

g. Are you likely to suffer from nerves? How will you relax?

h. Have you memorised the first two minutes so that you appear relaxed?

i. Does the chairperson know how to introduce you?

j. Is the audience clear about when to pose questions?

k. Have you anticipated the likely questions?

l. How will you promote interaction and discussion?

m. Will the audience be looking at print-outs of your material?

n. Do you have some ad-lib comments for the start and end?

o. Does your voice come across as confident?

p. How will you use the space available to connect with the audience?

q. Can you see a clock so that your time management is on track?

r. Who will you speak to immediately after to obtain some feedback?

4. Conversion

a. Did you need to debrief others from your presentation?

b. How well did you network with the audience before and after the presentation?

c. What will you send to the audience afterwards to establish a follow up?

d. Have you classified which of the contacts require urgent, later or no follow up?

e. Have you timetabled the various follow-up actions you agreed?

f. Have you integrated these leads with your overall sales management system?

Checklist and actions

- Have you developed a list of the key features and benefits that your firm/team delivers to clients in general?

- Do you always tailor the mix of features and benefits to meet the specific needs of each individual client?

- How often do you use the word "we" in your sales materials?

- How good are you at developing and delivering effective presentations in both informal and formal situations?

- How much time do you devote to setting out the aims, structure and key points of your presentations?

- Do you always use some form of visual aid to add interest and value to your audience?

- How interactive are your presentations?

- Do you ensure that you start with impact, stress the key points and end on a high?

- How do you assess the effectiveness of your presentations — and how do you improve them each time?

- How can you improve the value of the handouts that you provide at presentations?

Following up

Whether you are doing direct mail, attending a networking event, organising a seminar or participating in a local community event you will need to consider — in advance — how you will follow up any leads you generate. Too many people focus all their efforts on making contact at the event and return to the office with a stack of business cards and then fail, or forget, to follow up.

Often it is not due to a lack of motivation to make a follow-up call, but it is because insufficient attention was paid to determining and agreeing a reason to follow up when the contact was made. I mentioned above about the need to agree a low commitment follow up before you part — this means that the hard work is done because the person is expecting your call and knows why you are calling.

If you have already completed a lot of the sales legwork, then you might need to do a chase call. Here are some simple opening lines that you might use — with varying degrees of assertiveness, depending on the nature of your prospect:

- "I wanted to check that you received the letter."
- "Were there any other questions?"
- "Can I clarify or elaborate on any points?"
- "Did you need any further information?"

- "I wondered how you were getting on?"
- "Have you had a chance to think how you might proceed?"
- "We'd really like to do this work for you."
- "To get started ..."
- "I know that timing was important to you, when would you like us to..?"

Another interesting point is that many clients tell me that they were pleased to have met or had a discussion with a surveyor or agent and happy when a short confirmatory note arrived. But then they became busy and distracted and assumed, because the surveyor or agent did not make a follow-up call, that they were not interested. So get dialling.

It's in his pitch — pitching and tendering

You may end up doing a proactive pitch for someone's work (where you initiated the contact) or reacting to a formal tender document. But there are common elements and I have formulated a framework for organising how you respond to a tender to maximise the effectiveness of the huge amount of time you typically invest in these situations.

Competitive tendering framework

1. Pre-tender
   a. Marketing planning and positioning.
   b. Targeting and sales planning.
   c. Initiating contact and developing rapport.

2. Tender
   a. (If unsolicited) To bid or not to bid? (Assess the desire for the work and the chances of success.)
   b. Who will manage bid preparation?
   c. Who is on the team?
   d. What initial research and preparation is required?
   e. What do we do at initial client contact/meetings?
   f. What selling strategy do we adopt? (Why should they choose us?)
   g. What price shall we quote?
   h. How will we do the work if we win? (Work planning.)
   i. How do we write, produce and send the bid document?

j. How do we assess their reaction?
k. How do we prepare for the oral presentation?
l. How do we act at the presentation?
m. When and how do we ask for the business?

3. Post tender
    a. What have we learned?
        i.   from the team?
        ii. from the client?
    b. What do we do now?

I cannot explore all of these steps. So I will mention a few areas that those in the property industry usually find most useful.

Too many professionals are so happy to receive an invitation to tender, request for proposals or expression of interest that they fail to stop and think carefully about whether they really should be participating. It takes huge amounts of time to tender effectively, so why bother when there is really little chance of winning the work or the client or it is not suited to your overall strategy and strengths? So your first task is to understand why you have been asked to tender.

Ideally, you should have some procedures at your firm to guide people on how to respond to tenders. Some firms have a special risk assessment team. Others have a simple spreadsheet on their intranet that asks a series of questions and provides a rating on whether or not the pitch should be accepted. Some of the issues you might consider, when assessing a pitch opportunity, include:

*   Existing client?
*   Existing relationship?
*   Prestige client/case?
*   Relevant sector?
*   Skills/expertise needed?
*   Experience?
*   Partner commitment?
*   Even playing field?
*   Incumbent?
*   Reputation?
*   Conflicts? (now and in the future?)
*   Ability to pay?
*   Premium rates?
*   Potential for future work?

- Profitability?
- Location?
- Capacity of team?
- Notice/time available?
- Competitors involved?

What goes into the bid document? Larger institutions and public organisations will send you substantial questionnaires and forms to complete, so you will not need to worry about that. But when you develop your own materials, the following structure might help:

Bid document contents

1. Covering letter (in case it gets separated from the main document).
2. Title page.
3. Contents.
4. Executive summary.
5. Introduction/scene setting (show you have done your homework).
6. About the client (demonstrate your knowledge):
    i. issues, organisation, sector;
    ii. specific problem and solution sought; and
    iii. the various people who have helped/are involved.
7. The client's requirements.
8. Other issues identified. (Show you have considered the issue in a broader context, identify points that show you know your stuff, make them think that you are more knowledgeable than your competitors.)
9. Why (our firm)? (Use all your persuasion skills — remember the features and benefits.)
10. Relevant experience and expertise (credibility and positioning).
11. The team.
12. Project timetable.
13. How we will do the work. (Appear as if you are already on the case, present methodologies or systems that indicate your edge over the competition.)
14. (Our firm) client service.
15. Fees and terms.
16. The way forward.
17. Appendices.
    i. about (our firm);
    ii. about relevant practice areas;

iii. past cases/projects;
iv. client references;
v. CVs;
vi. relevant clients; and
vii. other information.

You should also check the material above on presentations, questioning and following up if you are shortlisted. One final word of advice, if you have to attend a pitch meeting "Check your ego at the door to avoid demonstrating your own brilliance at the expense of your clients' needs"!

Checklist and actions

- How consistent and effective are you at following up all marketing and sales activities?

- What systems and support exist to help you and your people follow up and pursue all sales opportunities?

- How much time does your firm devote to proactive pitching to target clients as opposed to reactive tendering in response to prospect and client requests?

- How do you measure the effectiveness of your tendering?

- Do you log the time that your people spend on tendering so you know the opportunity cost?

- Does your strategic pricing policy get used when developing price proposals for tenders?

- What systems are in place to assess the value and risk of tendering opportunities?

- How good are your pitch and tender documents and presentations?

- Do you properly debrief the team and the client following both successful and unsuccessful tenders?

Negotiating — competition or collaboration?

So most people in the property industry are brilliant negotiators — what more can I say? I suppose just to ask you to think about your negotiating style and whether you are generally achieve a "win-win" situation that is likely to position you for a productive long-term relationship rather than a "I win", which lands you this deal but may be a dead end in the future. When dealing with property professionals, I often think about the differences between a meeting, a discussion, an argument or a negotiation.

Michael Shea in his book *The influence sphere* outlines the personal characteristics of the good negotiator. How do you measure up?

1. Ability to read/assess opponents (how is your emotional intelligence and ability to read non-verbal communication?).
2. Know-how to assess strengths and weaknesses (if you are too self-satisfied you may overlook key issues and be unprepared).
3. Maximise and minimise aspects as appropriate.
4. Be a master at timing — know when to reveal information.
5. Be confident in using silence and prompting others to speak.
6. Have a straight face and the ability to bluff.
7. Use the threat of breakdown effectively.
8. Know how to distract, back off or put pressure on.
9. Be effective in applying psychological pressure.
10. Constantly question the opponent's position.

There is also much interesting material in the area of competitive verses co-operative negotiations. In a competitive situation you use power, argument, suspicion, hostility, deceit, resistance and posturing. In co-operative negotiations you use logic, search for the facts, openness and genuine problem-solving techniques to tackle real business objectives.

Closing the deal and handling objections

Whenever I am facilitating a sales training session — whether it is something as simple as networking skills, as day-to-day as tendering effectiveness or as detailed as complex selling situations in sophisticated markets — "closing" nearly always comes up as a question. This shows that some people still see selling in the professions all wrong.

Classical selling — where the aim is to obtain the business regardless of whether it is truly in the interests of the client and focused on the short-term transaction — taught about trial closing — where you "test the water" by exploring the likelihood of the client saying "Yes" before actually asking the question. And then there are all the old favourites for a formal close like:

- The conditional close "If we do x, will you be able to do y...?".
- The assumptive close "Well, that means we can get started on Monday ...".
- The alternative close "Shall we go with option A or option B?".
- The questioning close "So how would you like to proceed?".

But when you try to close a sale prematurely (ie, you have not fully uncovered the real needs of the client, you have not provided enough information, you have not addressed all their personal and business concerns, you have not established a trusting relationship, you have not demonstrated your credentials) an attempt to close will generate objections. But objections are not always to be feared — they are a signal that you need to provide more information or spend longer exploring the needs or potential solutions.

Objections are natural and inevitable — they may even be the prospect asking you to help them sell the idea to their colleagues. They almost always provide you with some insight into how they are thinking. At my very first sales training course, they presented a framework for dealing with objections:

- Pause and think.
- Clarify
  - Confirm your understanding of the problem.
  - Convert the objection into a question.
- Classify
  - Is it real (we do not have the money or authority), hidden (I am talking about price but I do not have the authority to make this decision), false (I am mentioning your reputation but really I cannot imagine the chief executive getting on with your senior partner), frivolous (My, you are expensive!) or a misunderstanding.
- Counter, answer or agree.
- Confirm
  - Check that all is well, that you have addressed all the issues

and can move on. Always end on a positive note, perhaps by reminding them of the overall purpose, value or benefits sought.

However, what the question about closing misses is that, in modern professional services, where the typical prospective commercial client relationship may have taken much contact over many months (even years) in order to gain trust, build rapport, explore needs, network through numerous decision makers, assess existing advisers, identify an appropriate opportunity and confirm the necessary authority and motivation there should be no need to "close". This is because you will have been working towards a business relationship over time and it is therefore a natural progression, rather than a stepchange driven by one question.

Words from an old master

Despite the many hundreds of books I have read and conferences and training courses I have attended, I still remember the huge impact on my own sales effectiveness made by one particular book. So if you have not read *How to win friends and influence people* by Dale Carnegie, I have taken the liberty of summarising his most important ideas — many of which I have already touched on.

- **Fundamental techniques in handling people**
  - Do not criticise, condemn or complain — people do not like negativity.
  - Always give honest and sincere appreciation — people like to be liked.
  - Arouse enthusiasm — an eager want.

- **Making people like you**
  - Become genuinely interested in other people (most salespeople fail because they try to get their prospects interested in them, rather than showing that they are interested in the prospects).
  - Smile.
  - Remember that a person's name is the most important thing that they can hear — so use it.
  - Be a good listener — and encourage others to talk about themselves.

- Talk in terms of the other person's interests — the less self-interest you express the better.
- Make the other person feel important and do it sincerely — comment on their ideas, ask their opinion and such like.

- **Win people to your way of thinking**
  - Arguments should be avoided at all costs.
  - Show respect for others' opinions (never say "you're wrong") but if you are wrong, admit it quickly and emphatically.
  - Begin in a friendly way and get the other person saying, "Yes, yes" immediately.
  - Let the other person do a great deal of the talking — at least 50%.
  - Let the other person feel that any idea is his/hers. Use empathy — try honestly to see things from the other person's point of view and be sympathetic with their ideas and desires.
  - Appeal to their nobler motives — for example, the good of the organisation.
  - Dramatise your ideas and throw down a challenge.

The trusted adviser

We talked about rapport and empathy above, but another vital component in a productive client relationship is trust. There is plenty of information about the importance of creating and maintaining trust. Some of the most valuable ideas include:

- "Trust is a means of reducing uncertainty in order that an effective relationship can develop." Gambetta (1988)

- "Trust refers to ideas concerning risk, power and dependency." Cousins and Stanwix (2001)

- "There are three outcomes from trust: satisfaction, perceived risk and continuity." Pavlou (2002)

- "Trust and commitment have significant impact on the creation of value." Ryssel (2004)

- "The components of trust are: reputation, familiarity/closeness, performance and accountability." Morrison and Firmstone (2000)

- "Trust is influenced by the duration of the relationship, the relative power of players, the presence of co-operation and environmental factors." Young and Wilkinson (1989)

[a] "The presence of commitment and trust leads to co-operative behaviour and this in turn is conducive to successful relationship marketing." Morgan and Hunt (1994)

David Maister, one of the greatest management thinkers for the professions, described in his book *The trusted adviser* a formula for assessing trust (see figure 6.13). There is also an online tool that you can use to assess your trustworthiness as perceived by your clients.

Figure 6.13: The TRUST equation

$$T = \frac{C + R + I}{S}$$

- Trustworthiness
- Credibility
- Reliability
- Intimacy
- Self-orientation

Source: David Maister

Checklist and actions

- What is your negotiation style? How could it be improved?
- Who are the best negotiators in your firm? What can you learn from them?
- How do you measure the success of your negotiators — the short-term deals or the long-term relationships?
- How well do you close deals?
- What are the most common objections you hear — and what are the best ways to avoid or manage them?
- How easy is it for clients to trust you and your people?

Many congratulations. You have hopefully improved the effectiveness of selling within your firm and won some new clients as a result. But now you have to keep the new clients and develop the relationship. Read the next chapter on how to make the most of your client relationships.

Large firm case study: Drivers Jonas LLP — selling

Drivers Jonas LLP (*www.driversjonas.com*) generates £98m income each year and employs 750 staff of which 110 are partners, and 25 are equity partners. Richard Crook, marketing and business development partner, runs a marketing team comprising 26 people (which includes eight research and information staff) to support the marketing and business development activities throughout Drivers Jonas's eight UK offices.

Property people like to think that they are good at selling and marketing but this presents a bit of a challenge. The transaction teams are generally quite good at it, whereas the others often loathe the idea. Marketing has come a long way here and the firm's culture has changed a lot to reflect this.

Motivating staff
Each surveyors' contribution to the marketing process — whether this is PR, winning new business, retaining existing clients or cross-selling — is considered as part of the appraisal process and is linked to their salary and bonus. It was a major driving force in achieving the cultural change we needed, as was ensuring we had the complete buy-in of Nick Shepherd, the managing partner.

Organisational change and multi-disciplinary teams
Part of the change was moving away from the classic 10 property divisions/service lines of agency, investment, property management, project management and so on to create a series of multi-disciplinary sector teams, such as retail, sport, residential, local and central government, health and education. We invite a team from a different sector to spend a week working together in the London office to focus on cross-selling. We have just had 12 people from the health team — building surveyors, L&T, valuations etc — from offices as far apart as Birmingham and Scotland. They have to produce a report at the end of the week for the managing partner. It's a great way to get them out of their comfort zone and to focus on the clients. It is amazing what they can achieve in just a week. But more importantly, these cross-disciplinary and cross-office sector teams gets groups of surveyors working and planning together to develop long-term marketing plans for the benefit of the sectors/clients they work in.

Thought leadership

The core of our marketing programme is the vast amount of seminars we present. This year we put on 110 seminars — that's two a week — from 12 person roundtables to 300 in an auditorium. We try to host as many as we can in our own offices. We publish a schedule for clients to use and cover topics as diverse as sustainability, green performance of buildings, PFI/PPP and town centres. Our regular crane survey (and yes, we actually send graduates out to count development sites) attracts huge interest from the media and many attendees to our events around the UK.

Sales training

We operate a four-day training course for every partner, associate partner and senior surveyor. They spend two days together learning about everything from the sales pipeline through client relationship development and a month later return to review progress for a day and a month later come back for a further day to discuss progress. Six months on they are able to book surgeries where they can pose any outstanding questions or talk through particularly challenging client situations. We also offer shorter courses on topics such as pitches, presentations and "schmoozing" (guidance for young surveyors on networking, body language, entering and leaving conversations, recording leads made at an evening, entertaining clients etc).

Tender process and support

We provide a lot of support for pitches. We have a pitch section on the intranet, which contains all the facts and figures, case studies and other information that they might require. Every tender has to go through an evaluation process before we decide to proceed — and each opportunity is given a mark out of 100 and has to be signed off by the unit partner. We then have a record of every pitch that is active within the firm — to help offices co-ordinate. Then, on the larger pitches, we often help prepare the document and the presentation and sometimes we even go along with them to the pitch. On key pitches, we organise a team debriefing to understand what we can learn for the future and we always ask clients for a debrief and most are more than happy to do so.

Measuring sales effectiveness

We measure effectiveness in a number of ways. We expect to reach the shortlist of 75% of those jobs we pitch for and we then expect to win 25% of those jobs. Generally, we meet those targets.

Obviously we record all fees but every month we also prepare a list of new clients won and on our new job forms we require surveyors to note down the source of the work — a seminar, a personal recommendation or an advert — and also whether the job was won in competition against other firms. Additionally, we record the nature of the client so we can see, for example, that 15% of new work in a particular month was from the health sector.

We also circulate a list each month of the firm's largest 250 clients and mark whether they are up or down on the previous month. This prompts some healthy competition between teams who then strive to develop their client relationships better than others — and that has to be in the firm and the client's best interests.

Advice for smaller firms

If I were to offer three pieces of advice to a smaller firm it would be:

a) Look after your existing clients first — too many firms simply chase after the new clients.

b) Engineer a culture where everyone in the firm has to buy-in to the concept that they all have a part to play in marketing and business development — from those making refreshments, through the secretaries and juniors and for all grades of surveying staff.

c) Promote your successes with good internal communication. People often forget to do this. But it is hugely motivating and it is important that everyone is aware of what the firm is doing and with whom, so that they can drop appropriate references into conversations with their clients, colleagues and friends.

Clientology and referrer relationships (relationship strategies)

Roadmap

- The importance of focusing on existing clients
- Touch points and intimacy — levels of client relationship management
- Delighting your clients — managing the client experience
- Managing expectations
- Take a walk on the client's side
- From the horse's mouth — designing your own client research programme
- Looking after the crown jewels — identifying and analysing your key clients
- Key account management programmes
- Getting under your skin — client intimacy
- The leader of the gang — relationship partner
- Beyond the call of duty — adding real value
- Key account plans
- Cross-selling
- Pigeon holed?
- I don't believe it! Managing complaints
- Stand and deliver — handing over your clients
- TQM, ISO, IiP — quality programmes
- You scratch my back — referrer management
- Let me entertain you

When I raise the subject of client care and relationship management with people in the property industry, I am often met with surprise. I know you provide a good technical service but I am interested in the philosophy, systems and procedures instilled in your people that ensure you pay as much structured and planned attention to retaining and developing your existing clients as you do to chasing down the next new client.

In effect, I am asking about the sales and relationship strategies deployed to keep and nurture existing relationships — both with clients and key intermediaries or referrers of work. Complacency is dangerous. And socialising (eg, drinking and playing of golf) is not the full solution.

The importance of focusing on existing clients

In the property world, it is commonplace to celebrate new client wins. No problem with that. However, in all the excitement of the cut and thrust of pitching for new business, it is easy to forget that it takes as much (sometimes more) effort to keep and develop existing clients as it does to win the new clients. We must remember that:

- It costs seven times more to win work from new clients.
- A 5% reduction in client loss can lead to a 50% increase in profits.
- A 10% increase in loyal clients can generate up to a 20% increase in profits.
- 90% recommendations arise from relationship issues.
- Satisfied existing clients generate income and greater profits.
- Service is the differentiating part of your brand.
- Higher-quality service justifies higher fees.
- Good service delivery is an integral part of a good relationship.

In a nutshell, looking after existing clients is as basic as the 3Rs — reputation, retention (repeat business) and recommendations/referrals.

Touchpoints and intimacy — levels of client relationship management

In the earlier section on strategic marketing (chapter 4 — section on branding), I addressed the importance of all the things beyond the delivery of technical expertise and managing the deal that make up the overall client experience of your firm or brand — and how these things have the potential to differentiate what you do from similar firms.

This is a big and rather "woolly" topic, so let us frame the different situations:

- **Client care**
  Ensure that everyone in the firm is focused on the clients' needs. In effect, this is every interaction that anyone in your firm has when in contact with a client. Whether this is answering the telephone, greeting them in reception, talking to them about an outstanding invoice or meeting them on site or in an agent's office. Every member of your staff in these situations is a brand ambassador, and what they do (and do not do) will reflect on your firm and their perception of its overall reputation or brand. One poor interaction can damage decades of careful nurturing.

- **Client development**
  Ensure that every agent and surveyor knows how to look after your clients in each day-to-day interaction for a particular transaction or assignment. The way they demonstrate knowledge of the client, the way they modify their approach to meet the particular requirements or style of that client and the way in which they deliver value to the client relationship at each interaction.

- **Key account management**
  In most firms, around 80% of the annual income will be generated by 20% of the clients (we call this the Pareto effect). Therefore, many firms have key account programmes to ensure that these special client or referrer relationships receive extra attention — usually by nominating a partner to take responsibility for overseeing all the interactions between that client and the firm.

Checklist and actions

- Consider how clients might react to the various touchpoints at your firm — when phones are answered at the switchboard or by agents or valuers, how clients are greeted in reception.

- Review the procedures and training that you have to help staff respond to clients in the correct way — consider the extent to which these procedures are followed in practice.

- Obtain a list of the firm's 50 largest clients — and consider how many of them receive particular care and attention from the partners.

Delighting your clients — managing the client experience

The first level of relationship management is ensuring that all members of the firm understand their critical role in ensuring that *all* communications with the client are conducted in a way that conveys the professionalism, client facing and other key values of your firm or brand.

Some firms use "mystery shoppers", who email, telephone and visit your firm to test out response and reaction. In some firms, the senior partners may do this on occasion. It may be necessary to get your support staff and more junior surveyors into a short (lunchtime) workshop to talk through what you expect of them, explain where and how the standards are set out (some firms may have these in their staff manual or in some form of client charter or even in their letters of engagement for new clients) and to hear what they have to say about how you can support them in delivering service excellence on a day-to-day basis.

Car parks often come under fire by clients and there is often little you can do to mitigate the problems of space — but clear instructions and expectations management before they visit your offices will help. Switchboard answering is another area that receives comments — so make sure you have the necessary resources to manage peak call times effectively and if you have to use an answerphone for outside hours outside then make sure there are clear procedures for taking and forwarding messages from the machine. Modern switchboards should contain software that enables you to track how many calls are received at different times of the day and how quickly they are answered. This means that careful monitoring can reveal weaknesses in the system and targets for improvement can be set.

I have spoken to clients who were upset that those on switchboard or answering phones for colleagues were not aware that they were major clients and/or regular callers — so check that everyone is aware of the major clients of each department, team or office so that they can be recognised and greeted appropriately.

Another problem area can be when initial enquiries come into the switchboard and the staff are either too busy or lack the relevant knowledge or information to ensure that they are directed to the right place. Training is one solution but with technology it is easy to set up "first points of contact" lists with key words to help telephone operators quickly divert callers to the right place. Some firms nominate particular individuals within each office or team to field those calls that require more careful exploration to identify the right person. Those in the marketing and business development teams must ensure that all staff are aware of any campaigns that are likely to generate calls from new clients. Also consider how such calls are to be logged and measured if campaign effectiveness is to be managed.

Beyond switchboard, each team should have a rota for answering calls and taking messages. There is nothing worse than having a phone ring and ring while it diverts from one handset to another and then finally ends up with a voicemail. Although, while holding for an architect's office recently, I was interested to hear a description of key services and projects, which made the wait more bearable. I remember well the senior partner of a larger property practice leading by example by always picking up a ringing phone as he walked through the offices. When I asked him how he managed to sound pleasant whenever he picked up a phone he said "I always imagine that the person on the other end of the line is just about to instruct me on a major new piece of work". You can imagine the impact of callers learning that their call and message was taken by the senior partner of a firm.

You should also ensure that your client-facing staff provide out of hours direct dial numbers or mobile phones if they suspect that their clients will wish to contact them early morning or in the evening. If you use voicemail make sure that people give callers the choice of leaving a message with a live person or voicemail and that staff update their outgoing messages frequently (and check them regularly).

Increasingly clients will use email to contact their advisers and expect answers almost instantaneously. This can be achieved with Blackberries and similar devices, although care must be taken not to offend clients whose meetings might be disturbed by constant calls. Again, good practice must be instilled in all staff in ensuring that their out of office assistants are used to provide up-to-date information about their whereabouts and likely return/response time and also to offer alternative contacts and numbers for urgent enquiries.

Reception areas are usually well maintained with attentive receptionists who keep papers and journals tidy and who offer visitors refreshments. Meeting rooms can be another area where the client experience is let down. Make sure these are cleared regularly and provide notepaper and pens as well as water and clean glasses.

Checklist and actions

- How do you regularly check on the quality of client interactions?

- How quickly are phones answered and messages actioned?

- Are there procedures (and training) for switchboard operators and receptionists?

- Does everyone who answers calls know where and how to transfer different types of calls?

- How are new enquiries measured and managed?

- How is your answerphone managed?

- Who looks after the car park, reception area and meeting rooms?

- What procedures (and training) exist for all staff on telephone, message and email management?

- Are staff encouraged to give direct dial and mobile numbers to clients? And answer their phones themselves?

Day-to-day client communications

Having ensured that communications for when clients come in contact with your firm are up to standard, you need to ensure that all communications with established clients who are conducting business with your firm are good.

There is plenty of information in chapter 6 on selling relating to the creation of rapport, trust and communication. This material applies equally in client relationship management situations, although obviously there should be a degree of rapport and trust between the original contacts at the client organisation and your firm.

One idea that is important though — as an adviser — is to recognise that different clients expect different styles of communication from their professional advisers — at times you might need to be prescriptive and directive and at other times you may need to be more like a coach or counsellor. Figure 7.1 shows the ways in which your style might need to vary with different individuals and different situations.

Figure 7.1: Communicating with clients

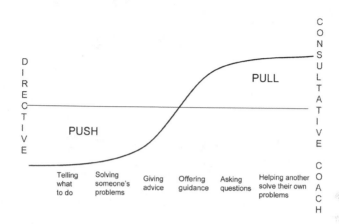

Managing expectations

One of the issues that is often revealed in client research is that clients' expectations are not managed as well as they could be. A little bit of communication goes a long way. When a client says something is urgent, do you know *exactly* what they mean? When you say that you will deal with something *immediately*, do they know when they can expect to see a result?

Client expectations can be managed. If we say something will be with them by 6pm it really must be there by 6pm, any later and they will be disappointed. Yet if you say it will be with them by 6pm and get it to them by 4pm then they will be delighted. You hold the key to your own destiny — and to happy clients!

This obvious but simple observation can have a dramatic effect on the quality of your client service. Perhaps you should think about incorporating it into your staff-training programme so that everyone understands the importance of delivering things on time and not making promises that cannot be met.

Figure 7.2: Managing client expectations

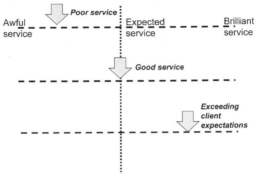

Source: Kim Tasso

Take a walk on the client's side

When we look at the general research material about what clients really want, there are no real surprises — except that they like empathy — the ability to see things from their point of view.

Clients' judging service quality

Berry and Parasuraman, 1991
- Reliability
- Tangibles
- Responsiveness
- Assurance
- Empathy

Harte and Dale, 1995 (accounting, architectural, engineering and consultancy firms)
- Timeliness
- Empathy
- Assurance
- Fees
- Tangibles
- Reliability

Empathy is: "The power of identifying oneself mentally with (and so fully comprehending) a person or object of contemplation", which itself is part of emotional intelligence ("The ability to perceive, integrate, understand and reflectively manage one's own feelings and other people's feelings" (Daniel Goleman). These are soft skills that are fundamental to winning new clients and retaining existing ones.

During the course of my consultancy work, I am in the fortunate position of often being able to ask clients what it is that they like and dislike about the service provided to them and what they would do to improve the service. Interestingly, while there are obvious differences between the clients of lawyers, accountants and surveyors and between particular firms — there are also many similarities.

Here are some positive comments that property clients have said to me:

- They are more responsive — they just get on with it. As soon as I ask them to act on a matter they do it.
- We choose them because of him personally, but also because of the back up he gets from the firm.
- We can always get hold of him — we have his home number, his mobile number. His secretary is absolutely superb at getting hold of him for us and keeping us in touch. He will ring back within the day or that same evening.
- They benefit from being a small, tight-knit team where everyone knows what's happening. If I ring with a question, there's always someone there who knows the answer.
- They seem genuinely concerned to help us as well as developing their own business.
- I can bounce ideas on him and he will tell me if he thinks we are going to do something wrong. He'd tell me when some consultants wouldn't.
- They keep in touch — we get weekly updates. They always let us know what is happening. They've picked up everything.
- You feel they are your friends and really working for you.
- The staff there seem to stay a while and the atmosphere is very nice — like they all get on.
- We have a very warm and close relationship.

Here are some more negative comments:

- I now have to think before I make a phone call to them because of the cost.

- He is very good but he needs someone else. He doesn't appear — or I haven't met — a really strong number two. One day he might be too busy to look after us.
- It's better to have a personal alert rather than junk mail.
- Now I know it's difficult but they really ought to be on time for appointments. It is very, very frustrating when they turn up late for meetings.
- It would be good if he could monitor us a bit closer. Some of the stuff needs to be watched very carefully.
- As they are so big things can get a bit diluted — they pass things on to others and they don't get done.
- It would be helpful if there was always someone in the office. Regularly we get the answerphone.
- They have a system whereas if it is a big job with a significant sign-off then you have to explain things twice and there is delay when it is signed by a senior person.

An inspector calls

Another simple but effective system to manage service quality and the client experience is where the senior partner (or head of department) has a system which prompts him or her to contact, each week, one or two of the firm's major clients to ask them about their satisfaction with the service provided. The senior partner will, of course, contact the relationship partner in advance of making the call to obtain any highlights or a briefing on any issues before making the call.

Along with providing an additional channel of communication to the client (there may be instances where the client is unwilling to reveal some information to his or her relationship partner or account manager), the call will demonstrate to the client that their business is valued by the firm. It also provides an opportunity for the senior partner to ask broader questions about the client's business aims and priorities that might be more difficult for the regular contact partner to ask.

Checklist and actions

- Are all staff trained in how to communicate effectively with clients in different situations?

- Are all staff aware of their role in managing client expectations?

- Do partners and senior staff set a good example in how they communicate with clients?

- Does the senior partner(s) take a personal interest in assessing client satisfaction on a regular basis?

- Do partners and staff exhibit sufficient empathy with clients and their needs?

- What would your clients say if asked about the positive and negative aspects of your service?

From the horse's mouth — designing your own client research programme

As well as gaining useful comments, a client research programme can also be designed to elicit scores against which you can measure your progress at improving client satisfaction. A sample "score sheet" is shown at the end of this chapter.

In fact, the chances are that the partners in your practice are confident that your clients are happy with the service and that there is no need for a "clientology" programme. Some information "from the horse's mouth" can often dispel their wrongly placed beliefs or, in a few rare circumstances, can be reassuring to senior management and provide valuable insights into your future marketing programmes.

Undertaking client research can be a difficult challenge. As a result, many firms use third parties to complete the interviews. Or perhaps you have a senior partner who is comfortable with interviewing clients and less likely to become defensive when clients start talking about past problems or other sensitive issues. It does need care when talking to clients, and it certainly needs a lot of care when providing feedback to the various partners about things that their clients may have said that indicate possible problem areas. Some people may want to shoot the messenger!

If you decide to conduct your own client research programme take some time to think it through as with any other critical project. As well as gaining information for a firm-wide and general client service project, the results can be used to provide a springboard against which individual client care or account management programmes can be established.

Designing and conducting a client research programme is a major project and you will need to reassure partners and staff that the information will be used sensitively and not as a witch hunt. Having said this, some firms do use client satisfaction scores (which they collect through an online system rather than through interviews) as an input to the appraisal and reward systems. But this is not recommended for the first time the exercise is conducted where you are more likely to want qualitative (subjective comments), rather than quantitative (ratings and numbers) information.

Illustrative project plan for a client research project

1. Agree the aims, scope and timing of the project.
2. Obtain internal agreement among the partners.
3. Confirm the specific information you seek.
4. Launch an internal communications programme to keep everyone informed.

5. Agree a questionnaire and method of interviewing.
6. Select a representative sample of clients to approach.
7. Prepare profiles/briefing notes — with contact details — for those clients.

8. Decide who will conduct the interviews.
     i.   If you decide to do this in-house:
          1. provide interviewing and note taking training.
     ii. If you decide to use an external adviser:
          1. identify suitable researchers and ask them to provide quotes;
          2. select the researcher that you feel most comfortable with; and
          3. invest time in familiarising the researcher with your firm/clients.

9. Set up the interviews.
10. Conduct each interview.

11. Thank each client for participating.
12. Brief each partner on their client's feedback.
13. Develop an account plan for each client.

14. Compile an overall report of all the findings.
15. Present the report and discuss it with partners.
16. Agree an action plan for firm-wide actions.
17. Communicate to all staff the implications and plans.

If — for cost or other reasons — you decide to undertake the client research internally, you should be aware of some of the pitfalls. Confidentiality is one challenge — I ensure that clients are given the opportunity for selected comments to remain unattributable. Also, I only provide detailed feedback from individual clients to the senior or managing partner and the relevant relationship partner — this is particularly important where clients have given feedback on specific individuals. The entire partnership might only see summarised and non-attributable comments that demonstrate the main trends and key issues arising from all the interviews.

An alternative to individual client interviews (either face-to-face or on the telephone) would be some form of group discussion — similar to a focus group. However, great care needs to be used to set up the right balance of clients, facilitate the discussion so that all members contribute and that the various topics are covered within the time frame allocated, manage complex group interactions and document the key points arising.

It is also important that the firm is seen to take action on any issues that are raised by clients. Otherwise the research and feedback process becomes undervalued. Some firms will feed back a summary of the research findings to all clients who participated, along with information regarding the changes and actions that the firm will adopt in order to address the main themes arising. Needless to say, all members of the firm should be alerted to the key issues and resultant actions — otherwise they will be caught off guard when their clients talk about them!

Checklist and actions

- Are the partners agreed that a client research project is needed?

- Are there sufficient funds and management time available?

- Have the aims and the scope of the project been agreed?

- Has a suitable researcher been identified?

- Are the questions clear and is the questionnaire robust?

- Have the right clients been selected?

- Have the clients been contacted and the exercise explained?

- Is there a timetable for booking, undertaking and writing up the interviews?

- Has the format for the final report been agreed?

- Have confidentiality issues been addressed?

- Have the relationship partners briefed the researcher?

- Is there sufficient time for the results to be analysed and discussed among the partnership?

- Has the project and the results of the exercise been communicated adequately among partners, participating clients and staff properly?

- Are there clear action plans — for individual clients and the firm — as a result of the research?

Looking after the crown jewels — identifying and analysing your key clients

Whether or not you decide to embark on a client research programme, you must focus some attention on your "best" clients. You may recall that we started a major client analysis in chapter 2 — this sections builds on that work.

Some of these "best" clients might be those that generate the highest income each year, some may be new clients with a huge potential for future growth and others may be those that you wish to nurture and retain because they are high profile, generate lots of referrals or for some other reason. Some firms may undertake an analysis of their key clients before embarking on a research programme, while other firms may undertake the research first to develop clear criteria for what they consider to be a "key client".

The starting point is to decide on your criteria and collect some information about their spend with your firm over the past few years and also about their spend across different teams or services. An example analysis is shown at the end of this chapter.

As well as providing you with a fast healthcheck on whether you are managing to retain and develop your most important clients, it will also provide some insight into the way in which you are effectively cross-selling your range of services within this important group of clients.

There are other types of analyses that you could use to understand better the nature of your existing clients and the appropriate strategies to manage them in a proactive way in the future. Furthermore, having a good grasp of your existing clients makes it easier to segment your market and target new clients (see chapter 4 on strategic marketing). Here are a few examples — from work by Ramachandran that some of my clients have found useful:

- Comparing value and potential value
    - Low maintenance (low value, low potential value).
    - Most growable (low value, high potential value).
    - Most valuable (high valuable, low potential value).
    - Supergrowth clients (high value, high potential value).

- Comparing satisfaction and loyalty
    - Hostages (high loyalty, low satisfaction).
    - Terrorists (low satisfaction, low loyalty).
    - Mercenaries (low loyalty, high satisfaction).
    - Apostles (high loyalty, high satisfaction).

- Loyalty
    - Monogamy.
    - Polygamy.
    - Promiscuous.

Keeping the important relationships with key clients front of mind within the firm is difficult. Some firms have a laminated sheet of paper showing the names of all the key clients (and the relevant relationship partner and other team members) that they circulate — although you need to be careful about confidentiality. Other firms use a dashboard that they show on partners' screens or as part of the intranet. The example below was developed using a simple spreadsheet and the partners became quite competitive about trying to increase the scores for "their" clients in the programme.

Table 7.1: Key client "dashboard"

Client	Production (fees)	Profit	Profile/ prestige	Propensity	Potential	Total
Client A	9	2	3	2	2	18
Client B	2	5	7	8	9	31
Client C	5	5	5	5	5	25
Client D	8	8	3	9	4	32

Source: Kim Tasso

Checklist and actions

- Do partners understand the aims and content of the key client programme?

- Have you agreed the criteria for the firm's key clients?

- Can the finance team provide the required information in the appropriate format?

- Have you selected the appropriate frameworks with which to analyse your key clients?

- Is there sufficient time, expertise and resources to implement the agreed plan?

Key account management programmes

As there is an enormous amount of information about selling (some of which is summarised in chapter 6 on selling), so there is an awful lot of information about account management. Many of the techniques and ideas are similar — for example, you plan how to develop the relationship during the pipeline phase of your sales strategy in much the same way as you develop an account plan on how to develop existing key client relationships.

> "Account management is an approach adopted by companies aimed at building a *portfolio of loyal key* clients by offering them, on a continuing basis, a service package tailored to their *individual needs*. To co-ordinate day-to-day interaction under the umbrella of a *long-term relationship*, companies typically form *dedicated teams* headed up by a *key client manager*. This special treatment has significant implications for organising structure, communications and managing expectations." Tony Millman (1995)

But I prefer the following definition, as it encompasses all the phases of client management that I mentioned in the introduction to the chapter:

> "A client-driven series of shared values, procedures and behaviours that, whilst differentiating the firm in a sustainable way yielding a competitive advantage, results in services tailored to specific client needs, high and improving client satisfaction and close and more productive and synergistic relationships with clients, which generate an increased level of work referral and greater profits." Kim Tasso (2001)

Again, in the textbooks we see a rather formal analysis of how key client relationships develop although the original ladder concept at the start of this book is more comfortable amongst the property professions (table 7.2).

Account management brings benefits to both the firm and to the clients (see above), but also clients benefit from account management:

- A one-stop, co-ordinated integrated and seamless service — clients do not need to concern themselves with the internal structure of the firm.
- A single point of contact — clients have a single point of contact through which all of their requests can be directed and therefore have easier and faster access.

Table 7.2: The stages of key account management

Development stage	Characteristics
Pre KAM	• Relationship distant and transaction focused. • Not all relationships potentially key. • Both parties assessing the others' potential. • Guarded information exchange.
Early KAM	• Exploring of possibilities for collaboration. • Tentative adaptations to provider's service/process. • Providers trying to build social relationships and trust.
Mid KAM	• Growth in trust and range of problems addressed. • Cross firm contact patterns increase. • Key account manager takes facilitating rather than lead role. • Increasing involvement of senior managers as potential for profitable collaboration increases.
Partnership KAM	• Buyer and seller closely aligned. • Senior managers from both sides closely involved. • Joint teams work on cost saving and quality issues.
Synergistic KAM	• Buyer and seller see themselves as a single entity creating joint value in the marketplace.
Uncoupled KAM	• Relationship ceases to be strategically important.

Source: Miller

• Well briefed advisers — all those working on their business are aware of the overall objectives and issues and are therefore better able to provide their specialist advice and services within the context of an overall framework.

Developing a key account programme is not a trivial exercise. It requires careful planning (with clear objectives so that you can measure your return on investment) and a number of separate projects to identify the clients on which to focus, research them to gain a deeper understanding of their business needs, training to help your partners and staff learn new relationship behaviours and perhaps even investment in systems to support the whole process.

Checklist and actions

- Have you considered the pros and cons of a formal key account management programme?

- Is there sufficient buy-in from all the partners?

- Do you have a clear project plan of what needs to happen and why and when to communicate the programme's aims and content?

- Are there sufficient resources in the technology, marketing and training teams to support the programme?

- Who has been given responsibility for designing, implementing and reporting back on progress?

Getting under your skin — client intimacy

The objectives of account management are as follows:

- **Really know the client's needs**
  To deliver a really good service requires an in-depth understanding of the client's business aims and procedures. To ensure that all the client's personnel, systems, industry and commercial issues and service requirements are known by the firm and by all those who work with the client — partners, assistants, trainees and secretaries. This enables all members of the client team to:
  (i) provide a consistently high level of service that is tailored to the client's particular requirements (which in turn will lead to higher client satisfaction); and
  (ii) identify areas where additional, innovative and added value services can be provided.

- **Increase the amount and range of work**
  To ensure that a substantial and increasing flow of work over an extended period of time — from a variety of sources within the client — are sent to various parts of the firm. In effect, account management enables the team to "grow" the client and protect it against approaches from your competitors. You need to gain a larger share of the client's wallet.

- **Differentiation and premium pricing**
  As we saw in earlier chapters, good account management enables you to differentiate the firm from its competitors by providing — consistently and in a way that is not easily copied — a different experience of team quality and responsiveness.

- **Co-ordination**
  To ensure that all activities across the firm in relation to a particular client are co-ordinated and viewed from an holistic point of view whether it is concerned with marketing, business development, entertaining, work processing or billing. The account management team ensures that the firm adopts a co-ordinated and integrated approach to the development of the relationship.

A starting point would be to prepare a relationship map so that you note down all the different people at your client organisation and understand their position in the decision-making unit (see chapter 6 on selling) and within the organisation generally in terms of their seniority.

Figure 7.3: Relationship mapping

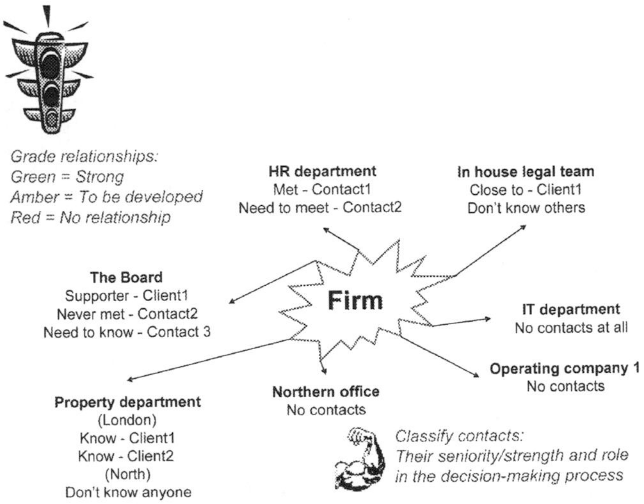

Grade relationships:
Green = Strong
Amber = To be developed
Red = No relationship

HR department
Met - Contact1
Need to meet - Contact2

In house legal team
Close to - Client1
Don't know others

The Board
Supporter - Client1
Never met - Contact2
Need to know - Contact 3

Firm

IT department
No contacts at all

Property department
(London)
Know - Client1
Know - Client2
(North)
Don't know anyone

Northern office
No contacts

Operating company 1
No contacts

Classify contacts:
Their seniority/strength and role
in the decision-making process

Source: Kim Tasso

This analysis should reveal where you need to develop further contacts or use your existing contacts to get in touch with other decision makers or users within the organisation. This can take some time, so you may need a plan to remind you of what needs to be done over time.

By analysing all of the contacts at the client organisation, you should also be able to overlay the people within your firm who also have dealings with the client. At this stage, you can identify whether you have a bow tie (all communications channelled through one person at the firm) or a diamond relationship (see figure 7.4).

Figure 7.4: Bow tie and diamond relationships

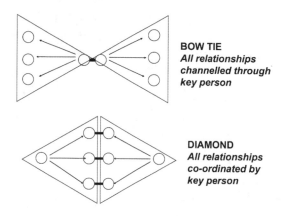

BOW TIE
*All relationships
channelled through
key person*

DIAMOND
*All relationships
co-ordinated by
key person*

Many firms will have bow tie relationships but these are at risk because if anything happens to the key person at your firm or at the client organisation it is likely that the entire business relationship will be put out to tender. Therefore, try to encourage people at all levels of seniority to establish close relationships with their counterparts at the client. But the relationship partner must take responsibility for ensuring that there is good communication and co-ordination across both organisations.

Once you have a grip on the relationships, you need to turn your attention to gaining a broader understanding of the client organisation. This means getting hold of their accounts, doing a detailed review of their website and publications, reading recent press coverage or comments from their trade press and even taking a look at their competitors.

This may feel a little uncomfortable as you are stepping outside of your "property" area of expertise, but it is important if you are to identify those issues and aims at your client organisation that might suggest the need for additional property solutions. Furthermore, it will underline your position with the client as someone who can provide useful ideas and input across a range of commercial issues.

The leader of the gang — relationship partner

While most clients will have a client partner, this is often the person who first introduced the client to the firm rather than the person who is best placed to manage the relationship on an ongoing basis. It is important to remember that the skills required to be a superb technical expert or agent are not the same as those for being a great relationship manager. In fact, in one property firm I worked with we identified a major strategic issue as being the lack of future "relationship partners" in the current partners and among the most senior associates.

Table 7.3: Comparing the skills of a technical specialist and relationship manager

Agent/specialist	Account manager
Transaction/case focus	Relationship focus
Fees for self/team	Fees for the firm
Narrow focus	Broad focus
Contact as needed during transaction	Regular contact at all times
Few, specialist contacts at client	Many, broad contacts
Output focus — speak and advise	Input focus — ask and listen
Short term and immediate	Long term and future
Technical	Commercial
Expert/specialist	Generalist

Source: Kim Tasso

It is clear that relationship partners have different outlooks and approaches to regular agents or surveyors — so what are the particular qualities that they need?

- Personal qualities
    - Integrity.
    - Adaptable (professionally and personally) and responsive to client needs.
    - Resilience/persistent.
    - Selling/negotiating.
    - "Likeability".
    - Keen and enthusiastic and prepared to think like the client so that they can be proactive in identifying opportunities.

- Subject knowledge
    - Specific professional service expertise/knowledge.
    - Understanding of the business environment and markets of clients.
    - Financial/budgetary.
    - Legal/compliance.
    - Computer literacy.

- Firm knowledge
    - Understanding of breadth of firm's knowledge base.
    - Experience of all the firm's systems and capabilities.
    - Familiarity with all the firm's services and expertise.
    - Helicopter view of entire client base.

- Thinking skills
    - Empathy/client's point of view.
    - Creativity and flexibility.
    - Strategic planning and thinking.
    - "Boundary spanning".

- Managerial skills
    - Focused on the end results of any professional advice — efficiency and effectiveness.
    - Communication, including listening and persuading.
    - People management and leadership.
    - Credibility — boardroom to post room.
    - Administrative/organisation.

- Client and professional skills
  - Understand client's business, aims, plans and market dynamic.
  - Ability to have an overview of the client's entire range of professional needs and how they interact.

It may be that you need to agree other key members of the client team — some responsible for communication and co-ordination, some responsible for managing specific jobs or specialist advice and others to maintain contact with those at all levels of the organisation. You may even need some administrative help in keeping your databases and internal information about the client up to date. You may want to consider who has worked with the client in the past as well as those involved with the client at present. You will need to consider which personalities work best with different individuals at the client organisation. Also, you should consider your future plans for that client — with regards to cross-selling — and identify who the best people from different teams and/or offices you should introduce.

What is it that the relationship partner needs to do beyond forming the appropriate client team? There is the analysis mentioned above and the account planning mentioned below. But there are also a series of day-to-day activities that must be considered.

Checklist and actions

- Have you identified the partners and senior associates who are best suited to the relationship partner role?

- Have you identified those clients where you are keen to take some action — and are your expectations of the results explicit?

- Do you have the resources and systems to obtain and manage the relevant information about the chosen clients?

- Are your staff organised into client teams with clear instructions on what they must do and incentives to devote time to relationship management activities?

- What training and coaching support are you providing to relationship partners and client teams?

- How will you want progress reported and how will you monitor/measure progress?

Day-to-day relationship management activities

In addition to managing the entire client team, undertaking analyses so that a clear understanding of the client is developed within that team and developing and implementing a client plan, the relationship partner has a range of other duties to perform. For example:

Internal/knowledge systems management
- Ensure that the client's policies, procedures, systems, best practice and preferences are documented and communicated to the entire client team.
- Ensure that the client's personnel and their interests are correctly recorded on the firm's database and marketing systems.
- Organise regular training for members of the client team to ensure that they are up to date with developments in the client's industry or business sector.

Team management and communication
- Schedule and prepare agendas and minutes of regular client team review meetings so that the service to the client is constantly being checked and improved.
- Introduce other members of the firm to the client and vice versa.

Satisfaction
- Assess client satisfaction in an ad-hoc informal way as well as part of more structured client service reviews.
- Provide a central point for any issues or concerns to be raised and deal with any complaints.

Financial
- Review all work in progress and bills before they are sent out to ensure that they are in line with quotes and agreed frameworks.
- Set targets for the fees (and profits) that are to be generated with that client.
- Check that bills are paid in a timely manner.

Progress reporting
- Organise regular formal and informal reviews with the client.
- Maintain systems so that the client receives regular progress reports on all outstanding matters and in the manner in which they have indicated.

Communication and co-ordination

- Maintain records on all work and communication between the client and the firm so that efforts can be co-ordinated and streamlined. Identify any areas where things might "slip between" different departments or offices.
- Update the client on relevant developments at the firm — key staff changes, new services, office or technology changes and also on changes within the team or unusual bills. This ensures that clients receive a "no surprises" service — one of the things that often comes up in feedback sessions.
- Some relationship partners develop one-page directories of all the team members and their contact details for the client — to facilitate easy and fast access to the relevant specialist while others manage the client's extranet (see below).

Business development

- Manage internal referrals to and from the client organisation.
- Scan the trade and industry media for the client and contact them where they may find it valuable to be aware of opportunities.
- Identify business opportunities for the client organisation.
- Be present at any semi-formal or social events where the client's senior personnel may be present to act as a good ambassador for the firm and to thank them for their continued business.

Furthermore, the relationship partner should be attuned to (and alert other members of the team) to the various political and other personal issues within the client organisation.

Table 7.4: Client politics

Rising star	**Fallen star**
Rapidly climbing	For old time's sake
Via a mentor	Promoted to incompetence level
Trusted by top guns	Limited influence
Gets information quickly	Bypassed on decisions
The fox	**The foot soldier**
Sly, moves quickly	One of many
Hard to identify — covert operations	May be aspiring
Good network — connected to top	Enlist support
Acts in own interests	
Rarely trustworthy	

Beyond the call of duty — adding real value

Client research reveals that they rate highly those relationships where they feel that their advisers go the extra mile or beyond the call of duty to add value to the relationships. So a key element of the relationship partner's role (and others on the team) is to identify ways in which real value can be added to the relationship.

By adopting the structured approach to key account management outlined above, it should be easier to identify opportunities to do this. But there are other ways. Many firms will provide "free" research reports, seminars or briefings to clients on topics of particular interest so that their staff receive the necessary continuing education points and remain up to date with new developments.

For example, in one property practice there was a partner who would — every couple of months — set aside 60 minutes in which to generate 60 ideas for his client that would help them in their business. He would spend a little time whittling down the list before running through it with his client. The client appreciated the time and thought that the adviser gave to his business and the conversations generally helped the adviser gain a deeper understanding of the workings of the client organisation and also of those issues that were mission critical. On more than one occasion, the client adopted some of the ideas generated, so real value was added.

In another firm, the relationship partner took responsibility for identifying what information and services her client would find most useful in a private extranet. This included access to the time/work in progress records, the ability to question particular aspects of the transaction management system that was in operation, real time access to key documents that were being developed, the deeds and lease database, key correspondence and a shared online workspace that contained instant messaging and group email facilities.

Sometimes you add value outside the business context. For example, another adviser was aware that a client was having difficulty with a particular situation in his private life. He offered to introduce the client to a friend who might be able to help with that problem and the client was extremely grateful. That gratitude cemented his loyalty to that adviser and the firm for many, many years.

Key account plans

Having gone through the analysis of your key clients and perhaps even a research exercise, as well as the allocation of responsibilities for managing the relationships and planned out what you intend to do, it would be good to capture the past information as well as the agreed actions into an account plan.

This will allow the relevant people in the firm to keep in touch with developments at the key clients and it will also provide a good source of information for informal discussions at partner meetings when people are looking for cross-selling opportunities.

Plans do not have to be long or elaborate. In fact, the simpler that they are the better. Remember too that, as with any planning task, the journey is often more important than the destination.

Where are we now?
- Client partner, account manager, team, succession.
- Map of contacts.
- History of relationship.
- Key items of correspondence.
- Background information (industry, company).
- Services used — opportunities/threats.

Where do we want to be?
- Objectives and strategy.

How will we get there?
- Messages and tactics.
- Team members.
- Innovative and new services or ways of working together.
- Contact schedule (formal and informal).
- Time-specific action plan.

There are many different ways to summarise a key client plan but an example plan outline is shown at the end of this chapter.

Database dilemmas

The first step in growing your business must be to ensure that you retain your existing clients and this requires you to have good information about your clients, the services they use and the people that you know at their organisations.

Without a shared database — to which everyone contributes their knowledge and information about clients — there can be no firm-wide client management programme. If you have read this far in the chapter without looking long and hard at the behaviours that prevent you maintaining a comprehensive, reliable and up-to-date database of clients, contacts and referrers, then you are probably wasting your time.

But it is easy to say, "have a shared database". There are many issues along the way. Do you include all clients? Do you include all information? Do you manage the database centrally or let everyone loose on it? How do you comply with Data Protection Act and privacy rules? Do you integrate with your accounts system or your property systems or do you have a standalone system? Do you have a separate system for your key clients? Do you keep information about prospects, targets, contacts and referrers in the same system? Just how much of an IT expert do you have to be?

These are tricky subjects and I do not have the space to do the topic justice. Suffice to say that you need to start with the strategic aims first — and look at what information you need, and then the behaviours needed to get that information — before you get embroiled in which technology platform to adopt. When you reach that stage — call in the technology experts.

Cross-selling

First, you should consider whether it is in the client's interests to cross-buy as some clients (usually the larger and more sophisticated ones) will have a "horse's for courses" approach to purchasing whereas smaller clients will appreciate the "one-stop shop" approach. A simple analysis may reveal that it is fruitless to try and cross-sell to a particular client, in which case you can direct your efforts elsewhere.

Benefits of cross-buying
* Convenient to obtain all services from one firm.
* Cheaper when bulk bought services receive a discount.

- Established relationships with people.
- Know and trust the firm and its people.
- Less time briefing other advisers.
- Less management input in co-ordinating different advisers.
- Reduced risk and uncertainty in selecting and using other advisers.

Reasons not to cross-buy
- Do not want all your eggs in one basket.
- Different directors have autonomy in their choice of advisers.
- Loyalty and long-standing relationships with other advisers.
- Need specialists in different areas from different firms.
- Keep multiple advisers as a matter of policy.
- Multiple advisers provide greater leverage on price negotiations.
- Other projects may distract my adviser.
- My current adviser is too busy serving my needs.
- Not aware of the other services provided by the firm.
- Never met any one from different teams at the firm.
- Inertia or too much effort to change advisers.
- Perfectly happy with other advisers.
- Not in my interest to promote my advisers to other parts of the organisation.

Checklist and actions

- Are relationship partners clear what they should be doing?

- Is there sufficient incentive for relationship partners to undertake their duties?

- How are you adding real value to your key client relationships?

- Have you outlined what you expect account plans to contain and indicated how they will be monitored?

- Are there clear and realistic expectations for what you expect any cross-selling programmes to achieve?

Pigeon holed?

Often, firms try valiantly to cross-sell their services without success and wonder why they have failed. Sometimes, it is important to explore the client's view of your organisation (perhaps as part of a research study mentioned above), as it could be because they have pigeon holed your organisation as being expert in only one field. The following framework has been useful to some clients to think through how the client sees you.

Figure 7.5: Client perception ladder

Level 4 — **Business partner**
(eg. manages all our property issues, is really close to the Board and we seek their views on everything at an early stage)

Level 3 — **Value-adder**
(eg. managing agents but with good ideas on increasing investment value and minimising tenant support)

Level 2 — **Product/service provider**
(eg. the best at valuations)

Level 1 — **Commodity broker**
(eg. manages office lettings)

While you can use good sales techniques to cross-sell to existing clients, it is probably better to do so from the context of a proper account management programme, so that you are sure that the additional services that you are promoting is in their interest and matches against their policies and present needs. You could even use the client perspective as part of your account planning to help you reposition your firm in your client's eyes.

I don't believe it! Managing complaints

Your professional body (or even your professional indemnity insurer) may have regulations indicating the type of systems that must be operated to manage client complaints. But it is good practice to have these systems in place — and training to support them — even if they don't.

While many professionals will view complaints as something to fear, in reality they are a valuable source of information from clients who are often keen to see the firm improve its service in some way. Often, dissatisfied clients will simply not instruct you again so those that take the opportunity to voice their concerns are usually quite loyal to the firm. There is significant evidence to suggest that a client who has complained and received a speedy and satisfactory response will become more loyal (and more profitable) in the future. So look at complaints as an opportunity.

It may be that you need to provide some training to partners and staff in how to manage people when emotions are high. Usually, clients will simply want their concerns to be heard and accepted. Listening patiently and acknowledging the client's unhappiness is a first step. Asking the client what they expect to happen to resolve the situation is another good strategy.

Avoid becoming defensive and offering "our version of events" at the outset, as this is likely to increase the client's irritation rather than improve matters. Ensure that the client knows what you intend to do and what the next stages will be. Provide a clear timeframe — even if it means that you have to hand over matters to others within the firm.

Always check with the client that they are satisfied with the outcome of any investigation or action and ensure that they are aware of what they should do if they wish to escalate the matter. There should be clear procedures in any case on what to do in problem situations from the professional negligence perspective.

Stand and deliver — handing over your clients

Within your key client account programme, you may hit difficulties when trying to hand over clients from one relationship partner to another. There are several reasons why you might need to hand over some or all of your clients:

- You may be taking a different role within the firm.
- You may need to use your time differently.
- There may be a personality clash between you and key members of the client organisation.
- You may feel that others in the firm have a stronger chance of developing the work flow from that client.
- You may wish to help a younger member of the firm build their client base/portfolio.
- It is just part of the natural development in an organisation where clients are transferred from senior people to the junior people as part of the day-to-day succession planning.
- You are about to retire.

Although the principal of sharing and transferring clients is a sound one and in line with a philosophy that "clients belong to the firm rather than an individual", it is fraught with dangers.

Clients may enjoy and value the relationship with their existing adviser and resent the imposed change. They may feel that their business is not valued and they are being "fobbed off" onto a younger and less experienced adviser. They may feel the organisation is growing too big and depersonalised. Whatever their feelings, attempting to hand them over may prompt them to look at the relationship overall and reassess it — possibly even seeking information from competitors.

If you are using a proper account management system, the problems and dangers are minimised as the client will already be familiar with several different people at the firm and therefore "handover" is no more than a change of emphasis on who they make most daily contact with, rather than a complete change.

However, the account management and other recommended long-term routes (see below) to handing over clients require a long-term view of client relationships and if this approach has not been embraced the ability to hand over clients without damaging the relationship is much reduced.

Long-term routes

- **Phased approach**
  Ideally, such a change should be planned in advance and implemented over a period of time. For example, the main contact should take a younger colleague along to meetings (shadowing) so that the client becomes accustomed to meeting the individual,

has time to get to know them and build rapport and trust. As time goes on, it becomes easier to have the younger colleague taking the lead on more occasions, so there is a gradual shift rather than a step change.

- **Policy**
  The firm should have a policy that all major clients should have reasonably regular contact with at least two members of the firm. For example, it should be policy (and enforced) that within three months of a new client relationship becoming established, a second member of the firm should be introduced to the client and start to form a relationship. In addition to reducing the need for step change handovers, this is also a good insurance policy for the firm to retain clients when individual consultants move out of the organisation.

- **Grading clients**
  Where all clients are to be handed over, the referring adviser should produce a list of all the clients to be handed over and to grade them according to the extent to which they will be at risk when handed over. General strategies can then be deployed for each "grade" of client and the majority of the effort can be concentrated on those clients who are perceived to need the greatest care during handover.

- **Peer matching/succession planning**
  A long-term approach would be to ensure that junior members of your team establish and develop close relationships with those people at the client organisation who are less senior than your main contact. This means that over time, as your primary contacts are promoted on or move out of the organisation, the next level of management takes the helm and their relationship with the firm is with those advisers at the next level down. This is an intrinsic part of the long-term account management approach.

Short-term approaches

- **Additional role**
  You could present the handover as an additional (non-charged) service. The existing main contact becomes the "account manager" — available for the most senior members of the client organisation to contact as required and adopting a more supervisory and quality

control role (ie, becomes closer to the client's needs) over the newer, younger consultants, who adopt a more day-to-day approach.

- **Quality/satisfaction audit**
  You might offer to undertake — at no charge to the client and with the specific aim of identifying ways to improve the service to the client — an audit. You could use a structured interview technique for an in-depth, face-to-face meeting undertaken by an external researcher, a fellow (objective) senior colleagues or yourself. Typically, during the course of the interview, the client will indicate additional needs or changes they require to the relationship and/or service delivery method. Once the client has identified the need for change, you effectively have a mandate to implement a change in the key point of contact.

- **Different service**
  You should assess the client's overall needs for advice and identify areas where you are currently unable to provide the necessary advice or resource. This leads to an opening to introduce another member of your team into the client organisation.

- **Events**
  Another way to introduce clients to other members of the firm is to invite them to an informal event — such as a workshop or a corporate entertaining function — where they can meet a selection of the firm's senior advisers. It should then be possible to observe where the client feels the greatest rapport and, in effect, let the client drive the handover.

- **Creation of temporary arrangements**
  Another method is to generate the need to provide a temporary alternative arrangement. For example, if the main contact is going on a longer holiday, taking a sabbatical to undertake research or some other "worthwhile" reason. This reduces the risk, offers clients the prospect of the resumption of normal service at some point in the future and allows a new relationship manager time to adjust while providing a back up (return to status quo) in the event of a problem.

- **More senior contact**
  You can appeal to the client's sense of importance by indicating that as they are so valuable to the firm, the firm feels it is

appropriate for a (the) most senior member of the firm to take a stronger role in the client relationship.

- **Fee incentive**
  Although not a recommended route, if a client has been expressing concern at the cost of service you might offer them a younger and less experienced member of the team to manage elements of the work at a lower cost than the present more senior adviser. The client then accepts a different point of contact as being the "price" they must pay for receiving lower fees.

Some clients will not handover

It may be that some clients simply will not tolerate a handover — regardless of the reasons or the approach adopted. In this case, the existing adviser will need to "front" all of the client interactions while using other members of the firm to do the "back room work". This requires a high degree of co-ordination and communication and may require an adjustment to the fees to take account of the additional/duplicate/supervisory work.

To ensure that handovers work, it is important that the fees and profits from each client are monitored closely so that any long-term decline in fees can be detected early and remedial action taken.

Whatever approach is adopted, it is vital that the referring adviser maintains some form of contact with the client. This may simply be a regular letter or call every three months or so to ask how things are going or a more in-depth, face-to-face discussion. In addition to showing the client that you retain an interest in their business, it provides an opportunity for you to assess ongoing satisfaction and the need to take any remedial action on the relationship.

Checklist and actions
- Does your firm's reputation make it conducive to promote the entire range of services to clients?

- What does your firm do to ensure that clients are aware of the full range of appropriate services?

- Is there a clear plan of which clients you expect to be handed over — to whom and by when?

- Have you set out a clear complaints-handling procedure and system?

TQM, ISO, IiP — quality programmes

People usually fall into two camps here. Those who see quality programmes as a fascination of bureaucratic jobsworths and those who see quality programmes as a vital management tool to set and maintain high standards and as an essential tool to help win public sector work and the approval of the buying professionals at large organisations.

I remember helping firms grapple with the paperwork involved in the original quality standards — BS5750 and the like and hearing comments that the systems really just ensured that you did things in the same consistent way all the time — even if it meant doing them badly. I was a bystander to the total quality management programmes which got firms to get their staff into teams to build commitment throughout the organisation on a quest for continual improvement. Also, I have sat by and commiserated with some first-class firms who lost out on major contracts because — despite being brilliant at what they do and the best placed advisers for that particular client — were unable to tick the externally accredited quality assurance certificate box.

So I am divided. Many of these systems require an unbelievable commitment in management time and resources to first achieve accreditation and then to keep it. Even so, in some markets you have no choice but to bite the bullet, whereas other firms can afford to leave it to one side as their market and client base does not require or value it.

However, you should seriously consider what you are trying to achieve and whether it is worth going down the quality standards route. Some firms have found that their existing management procedures and systems are so close to what is required that it takes little effort to go the extra little bit to achieve accreditation — then they obtain something that distinguishes them (almost) for nothing. Other firms have used the quality programme as a platform to take a root and branches look at their business and tackle all those things that have needed to be done for a very long time — but with a structured framework and a clear goal to motivate them.

Those firms who have invested a great deal in their human resources — through good recruitment, training, development and internal communications practices — have found that having something like Investors in People is more in fitting with their professional services roots and that it has provided a valued alternative for their clients to other quality programmes.

There was also one firm I worked with that operated a quality programme that they saw as a royal pain in the nether regions. They delegated all the paperwork and running around to one senior associate who toiled endlessly to get partners to provide the relevant information on time and in the right format. He never appeared to be demotivated by all the blanks on the reports that he was required to file on a regular basis. However, as we embarked on a strategic review we found that the reports he compiled gave a fantastic insight into the service standards of the various offices and different teams and also to the flows of work around the firm as well as the level of new instructions. Suddenly, the senior partners started to see that the processes that had been embedded in the practice had real value and started to use the quality management system as a serious business tool.

Love them or loathe them, anyone managing a professional service firm these days cannot afford to ignore them completely. I should add that the RICS management standards is an excellent template for how to ensure that your firm stacks up against the modern management practices at your peers' firms.

You scratch my back — referrer management

Many firms will find that a major source of their clients and work are intermediaries and other organisations that refer work to you. Many firms will receive recommendations and referrals from their existing clients. Others will have close relationships with other property professionals, lawyers, accountants and bankers that generate work opportunities.

These relationships need to be managed in the same proactive way as your key clients. Some firms have lists of their main referrers and maintain registers of referrals in and out of the firm (reciprocity registers) while others operate a less formal system, with each partner managing their own referrer relationships.

While some people think the best strategy is to have a broad and varied number of referrer relationships, others believe it is best to concentrate on just one or two. Either approach can work, although those at the start of their career may find it better to keep a wide number of contacts and more experienced professionals may find it more efficient to maintain close relationships with just one or two, who they know generate a reasonable level of referrals.

When selecting referrers on which to focus efforts, please consider the following:

1. Are they of a similar *size* to us? Referrals are more likely if they are. Sometimes, smaller firms may feel they can safely refer their clients for specialist expertise.
2. Do they share a *similar culture*? Are their people like us? Once a relationship between two individuals is established the similarity between others will become the focus.
3. Do we have any *mutual clients*? Working together is most likely to generate trust and provides a way for us to demonstrate our competence and way of working.
4. Are there several *areas of common* interest/work? The more areas there are, then the easier it will be to add others to the relationship and refer work to/from each other
5. Do they offer *competing services* (eg, financial services)?
6. Are we or they *leaders* in a particular area? You will be more attractive to those trying to get into markets where you are strong and less attractive to those where you are relative newcomers.
7. How many *other members* of the firm or their organisation have relationships? The more productive relationships you have, the more likely we will receive work in the future.
8. Are they currently close to *another firm* of agents or surveyors?

Manage your expectations — it can take time to develop a relationship with a referrer that has the necessary degree of trust and knowledge to make a work referral occur. Be realistic and accept that it may take many months (sometimes years) before all the time and energy you invest in a referrer relationship generates any work.

The basics

*   After meeting someone for the first time, make sure you do a *fast follow-up* (no more than two days after meeting) — a simple email, a copy of a relevant article or newsletter or the name of some contact they might found useful is all that it takes.
*   Add the name of your contact to your *database* to ensure that they receive regular newsletters from the firm. If your database allows it, note down their particular areas of interest so that they receive relevant information and invitations.
*   Advise your marketing team of the name of the individual and their organisation. Often, marketing will keep a list of all the firm's *referrer relationships* and can tell you which other members of the firm have links with that organisation. It also enables the firm to co-ordinate and measure all referrer relationships.

- You need a *personal filing and reminder system* to help you a) keep together all the information about the individuals and their referrer organisation and b) keep a log of past contact, topics discussed and work referred in or out. This does not have to be a computer-based system, but there must be some mechanism to remind you to stay in touch and to measure the effectiveness.

Plan

- *Analyse* your client and contact base, and your portfolio, to see where the majority of your work was referred from. Some people find that they generate most referrals from existing clients, and others from intermediaries, such as banks, lawyers and brokers. By analysing your data, you may find clues as to how to focus your referrer management programme in the future.
- There will, of course, be some referrers that you just meet, get on with and stay in contact with. However, you might need to seek out suitable firms proactively with whom to establish a relationship. Your marketing team should be able to help you review your portfolio, talk through your aims, put together a plan and help you research suitable organisations and opportunities to meet them.
- Be clear about the types of client and types of work *you are seeking*. The more specific you can be, the easier it will be for referrers to know when to refer people to you. Of course, the temptation is to say that you can do everything for everyone — but then all professionals say that — so make it easy and offer just one or two key services that will be of particular interest to the sorts of clients that particular referrer will deal with. They will come to associate you (and your firm) with a few key areas and it will be easier for them to remember.

Advisers to your key clients

- *Ask your clients* which intermediaries and professional advisers they use. Where you think there may be referrer potential, ask the client to facilitate an informal meeting. Alternatively, suggest that your client brings their other advisers along to a meeting or reception.

Grade your relationships

- While there is potential value in all referrer relationships it is necessary to *prioritise them* in order to focus your efforts on those most likely to generate work and to avoid spreading yourself too thinly. You may need to meet with your initial contact a few times and learn a little about their organisation before you can assign a high, medium or low priority to them. Then decide what you will do for the different levels of priority, for example, low-priority referrers may simply receive a copy of the annual newsletter, medium priority may receive a call from you three times a year and high priorities may actually meet with you several times over the year.

- Every six months or so you should set aside some time to *review* all your referrer relationships and consider which are proving valuable (and should be further developed) and which are going nowhere and should be given less time/attention. As part of the review you might identify gaps and areas where you need to develop new referrer relationships (see chapter 2 on planning).

Research

- Have a good look at their *website* and seek areas of common interest and/or mutual clients.
- If they are lawyers, check out *legal directories*, such as Legal 500 and Chambers, to assess their expertise and to see some of their most recent and important cases. For surveyors, use Propertymall website and/or RICS. For corporate finance referrers, there are various directories, such as Hemmington Scott and Insider Dealmaker.
- Ask your marketing team to prepare a *research pack* for you so that you have all the readily available information at your fingertips.
- Ask marketing to arrange for *news updates* about the organisation to be sent to you so that you have ongoing reasons to contact them and renew the acquaintance. Some firms subscribe to news services that do this automatically for you.
- Regularly *scan the relevant media* (legal, property, banking and local) to keep track of the major developments in the relevant referrer market so that you can ask them about their market and demonstrate your interest and knowledge.

Givers gain

- This is a principal used in networking. The idea is not to focus on what the referrer can do for you but to think about what *useful information and/or contacts* you can provide to the referrer — even if there is no benefit to yourself. By sending an email saying that you noticed an article that might be of interest to them or that you gave their details to someone who expressed an interest is all that it takes to trigger the "reciprocal" rule we looked at in persuasion — ie, they will feel indebted to you and feel compelled to repay the kindness in some small way.

After client or business meetings

* Instead of trying to arrange a lunch or dinner to get to know new contacts better, at the end of a business meeting (perhaps with a mutual client) arrange to go for a *quick/informal cup of coffee* or — if nearby — to see their offices. This is much lower commitment than a lunch or dinner and if you visit their offices there is a chance you will be introduced to their colleagues.

Regular contact

* Try to make contact on a regular basis — at least *every three months*. A short email or a telephone call to catch up will be sufficient.
* When reviewing your list of clients and contacts to *invite to seminars* or other events, consider whether any of your referrers might be interested in attending — and possibly encourage them to bring one or two of their clients or colleagues.

Third-party events

* Show that you are keen to be *invited to their events* — whether these are simply receptions or technical seminars. Ensure that you say hello to them when you attend and ask to be introduced to other members of their team.
* Ask if you can *bring a colleague* along to any of their events that you are invited to — that way you extend the number of relationships between the two organisations.
* *Identify which associations* — whether these are professional, commercial or technical — they belong to and assess whether you might accompany them as a guest or join as an associate member.
* If you are attending *key industry events* — such as an awards ceremony — call to see if they are attending and whether you might meet at the event.

Training

* All professional firms will have training programmes for their trainees and for their CPE (Continuing Professional Education). Ask them if there are any topics on which you could provide a lecture or workshop.

- Similarly, identify if there are any topics on which they could *provide a training session* for your trainees or young professionals. Or even more experienced members of the firm.
- An alternative might be to suggest that you *"exchange" young professionals* for a short period — say, a week — so that the relevant individuals develop a deeper understanding of each others' organisations and forge a wide range of relationships with numerous people at the referrer organisation.

Joint marketing

- Where opportunities exist, offer to provide a short presentation at one of *their client seminars*. Have a list of topics on which you are prepared to speak and which would integrate well with their speakers and be of interest to their clients to discuss with them.
- Invite them to *present a session* at one of your seminars or conferences.
- Offer to *write a short article* on some technical development in their client newsletters or on their website. Ask them to contribute an article to one of your newsletters.
- Suggest that you *co-write an article* for a trade publication reaching an audience of mutual interest. You may need some help from your marketing team to persuade a magazine to commission/ accept the article.

Lunch etc

- Once you have established some common ground it might be a good idea to *arrange a lunch* — but do not organise this too early in the relationship, as there may be insufficient to talk about and it is a high commitment of time. Make sure you have identified three or four topics of mutual interest to discuss and make sure you keep a note of the agreed actions — and follow up.
- Rather than a one-to-one lunch you might consider taking along *one or two other partners* and encourage the referrer to do the same. Again, prepare some agenda items and keep a note of agreed actions.
- Where the relationship is developing well and some clients/work have already materialised, it might be worth organising a lunch at your offices or theirs where around *six from each side* are present. You must select the people carefully to ensure there is likely to be areas of common interest. You are urged to agree an agenda in advance and take care with the seating plan. You might also

encourage each "team" to provide an introduction to their organisation and have each individual provide a short introduction to their area of practice and clients. To focus attention, you could set the goals of the lunch as "To find three good opportunities for each firm to pursue after this discussion".

- A further development of this idea is where two partners from your firm and two from the referrers organisation each bring around four other guests — a combination of clients and useful contacts and *a "roundtable" discussion* is arranged. A good roundtable will have a "guest of honour" who is prepared to provide a short opening speech and a good agenda (circulated in advance) or around six topics for discussion. A good chair is required to ensure that everyone contributes and that the debate keeps moving. Sometimes, a journalist is invited to attend with a view to writing about one or two of the issues discussed.

- An alternative to having a six-a-side lunch is to invite a group of their people from different levels of seniority (trainee to partner) to visit your offices at the end of the day for an *informal drinks reception*. This only works where a) there are a few people at your firm who know a few people at the referrer organisation or b) there is a clear list of topics to discuss or c) you have some form of speech or presentation at the start to help break the ice.

Hospitality

- Some sectors (eg, corporate finance, property) thrive on the social networks of corporate hospitality *entertaining*. Although many of these events are very expensive and time intense (eg, opera, major sports events) and should be used sparingly, there are plenty of other events that are less expensive and often more popular with some referrers (eg, go-karting, table football games, softball leagues).

- A development of this approach is to have the referrer organisation bring along one or two *well-known clients* to a hospitality event where you have taken along a couple of your clients — the key being to have clients who are likely to find it useful to network/meet with each other.

- Regardless of how informal your social/hospitality event, you should make sure that you make a *note of any actions/connections* that arise and follow them up.

Say "thank you"
- It may sound obvious but if they refer a client or piece of work please *express your thanks.*
- If a referral turns in to a client, without breaking confidences, *keep them informed* of how the work and/or relationship is progressing. Ask them if the client has commented on the service you are providing.

Remember, after all referrer contact to: debrief, note actions and follow-up fast.

Let me entertain you

A major part of clientology is the informal and social side of entertaining. Whether it is to cement the relationship out of hours, meet new people at their organisation or to introduce your colleagues or even other clients and intermediaries that they may find interesting, to say "thank you" for all the work that they have sent your way or just to have a chance to chat without the daily pressures.

In the property industry, most people appear to be real naturals at how to entertain effectively. Perhaps sometimes a little too much on occasions! However, the key thing to remember is to be sure that you understand your reasons for offering the entertainment, that it is the sort of entertainment that your client appreciates (not everyone wants to go hunting or watch rugby) and is allowed to accept (remember that many public sector workers face severe constraints on what they may accept and may have to disclose any hospitality on a public register) and that you do not end up enjoying yourself so much that you cause embarrassment or distress!

Checklist and actions
- Do you have appropriate resources and systems to monitor client and referrer entertainment programmes?

- Are partners and staff trained in how to behave at various types of hospitality?

- Do you have a list of the priority clients and referrers for different types of entertainment?

- Is there an annual plan of the key entertainment opportunities that links in to and supports the marketing and business development objectives?

Table 7.6: Illustrative client satisfaction scores from a simple research exercise

	S1	S2	S3	S4	S5	S6	S7	S8	S9	S10	S11	S12	S13	S14	S15	S16	S17	S18	S19	S20	Total	No Scored	Average
Working relationship																							
Accessibility/ease of contact	5.0	5.0	5.0	5.0	4.0	4.0	4.0	5.0	5.0	4.0	5.0	5.0	4.5	5.0	5.0	4.0	5.0	4.0	5.0	5.0	93.5	20	4.7
Speed of response	5.0	5.0	5.0	4.5	4.0	5.0	4.0	5.0	4.0	4.0	4.0	5.0	5.0	5.0	5.0	4.0	5.0	5.0	5.0	5.0	93.5	20	4.7
Regularity of contact	4.0	5.0	4.0	5.0	4.0	5.0	3.0	4.0	3.0	4.0	4.0	4.0	4.0		3.0	4.0	4.0	3.0	5.0	4.0	76.0	19	4.0
Flexibility in approach	5.0	4.0	5.0	5.0	4.0	5.0	5.0	4.0	4.0	4.0	5.0	5.0	4.0	5.0	5.0	5.0	5.0	4.0	5.0	5.0	93.0	20	4.7
Friendliness	5.0	5.0	5.0	5.0	4.0	5.0	5.0	5.0	5.0	5.0	5.0	5.0	5.0	5.0	5.0	5.0	5.0	4.0	5.0	5.0	98.0	20	4.9
Partner attention	5.0	4.0	5.0	5.0	4.0	5.0	4.5	5.0	5.0	5.0	5.0	4.5	5.0	5.0	5.0	4.5	4.5	4.0	5.0	5.0	94.5	20	4.7
Professional staff	5.0		5.0	5.0	4.0	5.0	5.0		3.5	2.0	4.0	5.0	5.0		4.0	4.0		3.0			56.5	14	4.0
Professionalism	4.5	3.0	5.0	5.0	4.0	4.0	5.0	4.0	3.0	5.0	4.0	5.0	4.0	5.0	5.0	4.0	5.0	4.0	4.0	4.0	86.5	20	4.3
Agreeing terms and fees	5.0	3.0	5.0	4.5	4.0	5.0	3.0	2.0	3.5	4.0	4.0	4.0	4.5	4.0	4.0	3.0	4.0	2.0	4.0	4.0	76.5	20	3.8
Offices and premises	4.0	5.0		4.0	4.0	4.5		3.5	4.0	3.0	3.0	4.5	4.0		4.0	3.0	4.0	4.0	4.0	3.5	65.0	16	4.1
Ideas and innovation	5.0	3.0		3.0	3.0	3.0	3.5	3.0	2.0	2.0	4.0	4.0	4.0		3.0	3.5	4.0	4.0	2.0	5.0	61.0	18	3.4
Use of technology	5.0	3.0	3.0	5.0	4.0	3.0		3.0	4.0	4.0	4.0	4.0	3.0		4.0	3.5	4.0	3.0	3.0	3.0	48.5	13	3.7
Technical legal expertise	5.0	4.0	4.0	4.0	4.0	5.0		5.0	4.0	5.0	5.0	5.0	5.0	5.0	4.0	4.0	5.0	4.0	3.5	3.0	83.5	19	4.4
Commercial knowledge	4.0	4.0	3.0	5.0	4.0	5.0	4.0	4.0	4.0	3.0	5.0	4.0	4.0	4.0		4.0	4.0	3.7	3.0	4.0	75.7	19	4.0
Sector knowledge	4.0	4.0	3.0	3.0	5.0	3.0	4.0	5.0	4.0	3.0	4.0	4.0	4.0		4.0	2.0	2.0	3.0	4.0	4.0	71.0	19	3.7
Services																							
Residential agency	5.0							3.0	3.5	4.5	4.0	5.0	5.0	5.0	5.0	4.0	5.0	3.0	5.0	3.0	78.0	18	4.3
Residential lettings		4.5	5.0	5.0	5.0	5.0		4.0	5.0		4.0	5.0	4.0		5.0	4.5	5.0	4.0	4.0	4.0	73.0	16	4.6
Commercial agency		4.5		5.0	5.0	4.0		4.0			3.5	3.0		5.0				3.0	3.0		34.0	8	4.3
Valuations											4.0	3.0	4.0	5.0	5.0	4.0	4.0	4.0	4.0		33.0	8	4.1
Rating			5.0	5.0									3.0						5.0	5.0	17.0	4	4.3
Corporate real estate		4.5		4.0	4.0						4.0	4.0	5.0		5.0			4.0		3.0	38.5	9	4.3
Industrial agency																3.0					3.0	1	3.0
Building consultancy						4.0															4.0	1	4.0

	S1	S2	S3	S4	S5	S6	S7	S8	S9	S10	S11	S12	S13	S14	S15	S16	S17	S18	S19	S20	Total	No Scored	Average
Overall																							
Technical expertise/ knowledge	5.0	4.0	4.0	5.0	4.0	5.0	4.0	5.0	4.0	4.0	4.5	5.0	5.0	5.0	4.0	4.0	4.0	4.0	4.0	4.0	87.5	20	4.4
Quality of service	5.0	4.5	5.0	4.5	4.0	4.0	4.0	4.0	5.0	5.0	5.0	4.0	5.0	4.0	4.0	4.0	4.0	4.0	4.0	5.0	88.0	20	4.4
Relationship	5.0	5.0	4.5	4.5	4.0	4.5	4.0	3.0	4.0	5.0	5.0	4.5	5.0	5.0	5.0	4.5	5.0	5.0	5.0	4.0	91.5	20	4.6
Reputation	4.0	4.5	3.5	3.0	3.0			4.0	4.0	2.0	4.0	5.0	4.0		3.0		5.0	3.0	4.0	3.5	59.5	16	3.7
Performance	5.0	4.5	5.0	5.0	3.0	4.0	4.0	4.0	4.0	4.0	4.5	5.0	4.0	5.0	4.0	4.0	4.0	3.5	4.0	4.5	87.0	20	4.4
Value for money	5.0	4.0	4.5	5.0	4.0			3.5	4.0	5.0	4.5	5.0	4.0	4.0	3.0	4.0	5.0	3.5	4.0	4.0	76.0	18	4.2
Total for client	99.5	94.0	90.5	119.0	102.0	99.0	66.0	96.0	95.5	86.5	113.0	116.5	111.0	82.0	98.0	85.0	105.5	88.7	95.5	99.5			
Number of scores from client	21	22	20	26	25	22	16	24	24	22	26	25	26	17	23	22	24	24	23	24			
Overall average for client	4.7	4.3	4.5	4.6	4.1	4.5	4.1	4.0	4.0	3.9	4.3	4.7	4.3	4.8	4.3	3.9	4.4	3.7	4.2	4.1			

Source: Kim Tasso

Table 7.7: Illustrative major client analysis (over £10,000 fees pa) and cross-selling matrix

Client name	Partner	Sector — business	2005	2006	2007	YTD 2008	Projected 2008	Agency	Management	Valuations	Consultancy	Projects	Notes
									CROSS SELLING MATRIX ANALYSIS 2007				
Accountancy heroes	B	Professions	111	222	333	155	333		333				International potential
Airline One	C	Transport	111	111	111	111	111	222		111			
Association	X	Charity	111	222	222	111	222						Overseas potential
Bank 2	D	Bank	111	222	444	333	555		222	222			
Big firm	Y	Professions — property	888	999	777	333	666	666				111	
Buy me	Z	Retail	0	999	0	111	555						
Connections Plc	T	Telecoms	666	555	444	111	333				444		
Consulting 2	X	Business services — marketing	999	222	888	888	999	888					Going in house?
Data company	G	Technology	555	555	999	888	999			888	111		
Dealership	G	Distributor — car dealership	111	222	111	222	333				111		
Designers Inc	B	Importers	111	111	222	222	333	222					At risk
Developer 4	Z	Property — development	0	333	0	111	444						
Eat well and be merry	Z	Retail	0	0	888	555	555	777				111	
Finance house 1	G	Finance	222	333	444	333	555		111	333			Financing work
Fund	B	Charity	444	444	444	444	888	444					
Get together today	F	Charity	111	0	0	111	222						
Give well	D	Charity	666	666	0	0	0						
Government	X	Government	888	888	999	777	888	999	222				Active research programme
Heating Contractors Ltd	B	Construction	222	111	222	111	555						
Help them	F	Charity	999	888	777	333	666	777					
High Street shop	F	Retail	0	111	222	333	555			222			
House builders plc	Y	Property — housebuilders	888	222	444	222	666	444					
Household stuff	T	Distributor	222	222	0	333	666						
Insolvency firm	F	Profession	0	111	333	222	555			333			
Lovely property company	X	Property	0	0	999	222	444	999					

CROSS SELLING MATRIX ANALYSIS 2007

Client name	Partner	Sector — business	2005	2006	2007	YTD 2008	Projected 2008	Agency	Management	Valuations	Consultancy	Projects	Notes
Manufacturer 2	D	Manufacturing	999	888	777	333	666				777		
Media Group 3	B	Media — publishing	555	666	444	0	0		444				Pitched and lost
Mega marketing company	X	Business services — marketing	888	888	888	888	888	888					
Partners United	G	Professions — property	333	222	333	111	333						
Property Co 4	Y	Property	222	666	888	444	999	888			222	111	
Publish and be damned	Y	Media — publishing	111	555	333	222	444			333			
School	D	Charity — education	111	111	111	111	222	111					
Secondary foreign bank	G	Bank	444	555	444	222	444		444				
Shop til you drop	A	Retail	888	0	0	444	444						On hold for present
Small charity 4	Y	Charity	333	555	555	111	444	555					
Systems 1	D	Technology	0	444	333	111	222			333			
Wired for sound	D	Telecoms	0	0	0	999	999						New for 2008!
Total			13,320	14,319	15,429	11,588	19,203	8,880	1,776	2,775	1,665	333	15,429

Source: Kim Tasso

Table 7.8: Key account plan — summary

About the client/prospect/organisation	
Name	
Established	
Address	
Telephone number	
Website	
Turnover	
Sector/business	
Employees	
Main location	
Other locations	
Auditors	
Bankers	
Solicitors	
Other advisers	

About the relationship	
Date relationship started	
Current spend with us	
Current spend with others	
Potential estimated spend	
Key achievements	
Key issues	

Account team	
Lead partner	
Relationship partner	
Other partner 1	
Other staff 1	
Other staff 2	
Key issues	

Account meeting schedule

Topic	Date	Location	Attendees

Opportunities

Key client issues	Our opportunity	Current status*	Potential fees	Key client decision maker	Our lead

* Exploring/identified/qualified/pitched/won/lost

Source: Kim Tasso

Table 7.9: Relationship map

Name	Role	Relationship influence	Power 1=Low 10=High	Allegiance	Desired allegiance	Our lead contact	Comments

Relationship influence:
Decision maker
Gatekeeper
Evaluator/buyer
Influencer
User
Sponsor
Anti-sponsor

Allegiance:
Unknown
Neutral
Challenging
Supporting
Advocate

Small firm case study: Spacelab — clientology

Spacelab is a multi-disciplinary architectural practice based in East London. The practice was formed in 2002 by Nathan Lonsdale and Andrew Budgen and has since gone on to win a number of prestigious awards. The client list currently boasts organisations such as Emap, Bauer, Virgin Management, WPP Group and Great Ormond Street Hospital. Spacelab currently employs 13 members of staff.

PR and publicity

To date the majority of Spacelab's work has come from word of mouth or referrals and recommendations from various friends and contacts. Good PR has also played a vital role in increasing the awareness of Spacelab as a brand and generating new business leads.

Spacelab's first big architectural break came in the form of "The Westlake House" featured on Channel 4's *Grand Designs* in 2003. At the time, the owner of the house worked in production and had connections to the programme.

Grand Designs generated a huge amount of publicity for Spacelab and continues to do so with reruns of the show and *Grand Designs Revisited*. Spacelab has also had a large number of enquiries as a direct result of the show, which led to their collaboration with Great Ormond Street Hospital.

Friends in high places

The biggest break for Spacelab as a company has come in the form of Emap. Spacelab was introduced to Emap by the head of property, a contact of Nathan and Andrew's from their previous place of work. Since completing its first project, Spacelab has gone on to work on a vast range of small and large-scale projects across Emap's property portfolio. Emap provided Spacelab with the backbone of its income until 24 months ago and remains a valued client. From the experience gained at Emap alongside the invaluable contacts made, Spacelab has gone on to win clients such as the WPP Group and Virgin Management; and is now working on their second project for each.

Brands, website and extranet

Two events have had a major impact on Spacelab. Rebranding the business in 2005 helped to shape the vision of Spacelab's future as well as representing and reinforcing Spacelab's message. During this time Spacelab also decided to recreate their website. This made it far more accessible to users as well as visually showcasing all of Spacelab's work and providing information on what Spacelab has to offer. The website has received great feedback from clients and has become a crucial marketing tool. An extranet for Spacelab's key clients was created to save clients' time when requesting drawings. Clients are able to log-in and download updated drawings without having to call the office. The information is there when they need it rather than having to wait.

In 2007, Spacelab decided to relocate to a more central location, enabling clients and visitors easier access to the office. The property was purchased as an empty shell and Andrew, with the help of the staff, created a space that met the needs of the business as well as being a creative environment to work in. It was an opportunity to show current and prospective clients how far Spacelab had come as well as boosting staff morale and satisfaction.

Keep the client in mind

Nathan and Andrew have always endeavoured to maintain open and honest relationships with their staff and clients and, as a result, many clients still remain good friends of theirs today. They also have an extremely low staff turnover. Nathan and Andrew's approach to client relationship management has always been "keep the client in mind". It is ensured that contact is made with every client at least once a week, whether it is by telephone or email. Clients and various stakeholders receive regular newsletters updating them with all Spacelab's projects and notifying them of website updates. Every Christmas, clients receive a Christmas card and Spacelab take them out for a meal to say thank you.

From appointment onwards, both Nathan and Andrew have worked closely with clients, holding numerous meetings and steering groups to help the process of creating their brief. Spacelab always strives to make sure the client has all the facts they need to make the right informed decisions. This is to make sure the client has what they truly need rather than what they perceive they need, even if this sometimes means telling the client they are wrong.

Ideas for the brief are discussed with members of staff from various business areas, as well as management, to gain a united view of what the staff want as well as giving staff a chance to put ideas to the board. Spacelab realises the importance of involving all staff and to give them confidence in the changes that are being made to their work environment and embrace them.

Networking

Socialising and networking have also played a vital part in the development and success of Spacelab. At least once a year Spacelab organises a party, usually at the site of a recently completed project (in 2007, it was based at Spacelab's new offices and in 2008, Great Ormond Street Hospitals' Roof Top Terrace), to allow clients a chance to view recent work. Wherever possible, staff attend industry events such as award ceremonies, seminars and conferences. Staff often take part in client "days out" whether they be a few rounds of golf or a charity fundraiser.

One year both Nathan and Andrew attended the RIBA's version of speed dating. People within the industry were given an opportunity to network and make new contacts in a social environment.

Nathan and Andrew realise the importance of these opportunities, ensuring that any contacts made are always followed up with meetings or informal drinks/ dinners. From the director's perspective, a more personable approach often sees more benefits. In any industry people will always prefer to work with people they know and trust, it is therefore worth building on these relationships.

Overall, the combination of good PR and Spacelab's personable approach has contributed to its success. Managing their client relationships and "going beyond the call of duty" has seen one-off clients develop into long lasting working relationships.

Talent, teams and tomorrow's leaders (people strategies)

Roadmap

- Who's in charge? The staff partner
- The experts — the human resources manager
- The way we do things round here — culture
- *Vive la différence* — celebrate people's differences
- Tantrums and time wasters — dealing with difficult people
- Time to talk? Internal communications
- Talent spotting — recruitment
- Keeping the saws sharp — training and development
- Encouragement through coaching and mentoring
- What did I do wrong? Appraisals and feedback
- Why should I? Motivation
- What's in it for me? Reward systems
- Rubbish! Dealing with underperformance
- Unsung heroes — support the support staff
- It's all in your mind — knowledge management
- Succession
- And don't forget ...

"Human resource management is a series of activities, which: first, enables working people and the organisation who use their skills to agree about the objectives and nature of their working relationship and, secondly, ensures that the agreement is fulfilled."
Torrington, Hall & Taylor

Whether you have just a couple of staff or hundreds, you will find that they take up the lion's share of your time and your costs. For your firm to move from the "small and growing rapidly" to the "sustainable and profitable growth" model we described earlier, you need to have a handle on the core human resource processes. While human resources management is a huge topic, there are a few key issues that all surveying practices must address:

- Talent — attracting, retaining and developing the right quality people.
- Teams — building effective teams, so that work is managed efficiently.
- Tomorrow's leaders — harnessing the power of the next generation and growing tomorrow's partners.

Who's in charge? The staff partner

Smaller firms may have a partner with specific responsibility for staff matters — the staff "champion". Some firms will have separated out the particular challenge of graduate recruitment and training and allocated this to a different partner. While everyone in a supervisory position will have a role and responsibility for human resources matters, you need someone to take charge and overall responsibility. As human resources is such a huge subject — and one that is vitally important to both the day-to-day smooth running of the firm and its long-term success — you should set out clearly what you expect from the person with ultimate responsibility for staff and talent matters.

Your business plan (see chapter 2) will have set out what you want the firm to achieve over the next five years and, as a result, what your partner and staff requirements are likely to be. Your business plan should also set out any key human resource issues that need to be addressed. One of the first tasks of the "staff champion" should, therefore, be to produce a firm-wide human resources plan showing what they will focus on — their objectives and their strategies — for ensuring that the human resources element of the business plan will be met.

If you are relatively new to modern management practices, you might also think about developing and agreeing some terms of reference for what you want your human resources or staff partner to tackle. And indicate the boundaries of their responsibility. Show when and how you expect them to consult with the wider partnership and when they can act on their own initiative. If you really want to ensure that they are successful — and not just the whipping boy — then you ought to allow them to attend some basic training on human resources management. Furthermore, allow them to apply some of what they learn when they return all motivated from their courses.

In addition, you cannot simply hand over an important responsibility to a partner and expect them to carry on as usual without access to additional resources — whether this is cash, their own time or support from other people. As such, you might have to modify their time-recording or fee-generating targets to allow them to devote sufficient time to do the task properly.

Checklist and actions

- Does your business plan indicate the firm's requirements for present and future human resources?

- Who has overall responsibility for staff matters at your firm?

- Have you articulated what you expect your staff partner to achieve and the limits of their authority?

- Has the person with responsibility for human resources matters had some appropriate training?

- Do you have a way of measuring the success of your staff partner?

- What resources will be available to the staff partner?

- How will you reward the partner with responsibility for staff matters?

The experts — the human resources manager

Many firms employ specialist human resources people. But very often these people (or person) are tasked with the day-to-day administration of staff matters — writing and placing recruitment ads, liaising with recruitment consultants, setting up interviews, trying to get appraisals done, inductions for new staff and organising cover when support staff are absent.

Their training and knowledge are deployed merely to manage the operational side of things — and they are not allowed to become involved in senior management discussions about the strategic aspects of human resources. Sometimes, being cash conscious, you employ an HR professional that is at such a junior level that they do not have the training, knowledge or experience to advise on strategic human resources matters.

Ideally, your HR specialist should be qualified — the Chartered Institute of Personnel and Development (CIPD) is the relevant professional organisation. If they are young, then consider supporting them morally and financially in achieving their professional qualifications — it will benefit the firm in the end. If they need to gain experience and "grow into" the role, then consider inviting them to join management discussions about such matters so that they start to gain a more strategic perspective on things.

If your budget does not stretch to employing a qualified HR person and you have serious talent issues, then consider whether it is worth calling in a consultant (see chapter 5 on tactical marketing to show you how not to get your fingers burned with consultants).

Checklist and actions

- Is there a clear job description for your in-house human resources expert?

- Are you clear about what you expect from your in-house human resources expert and what you want them to deliver?

- Is there a need to employ an external consultant to assist you with a critical human resources issue?

The way we do things round here — culture

A firm's culture is probably the main way in which it is differentiated in the marketplace — both in terms of the service it provides to clients and in terms of its attractiveness to potential recruits (the "employer brand"). In small firms, the culture is usually determined by the particular values, behaviours and style of the founders. Such is the power of strong role models, that those joining the firm will relate to and absorb these values.

However, when a firm experiences rapid growth or there is a merger, the culture can be diluted or even destroyed. While it is extremely hard to define culture generally or to describe the culture of a particular firm, its corruption or loss can have a devastating effect on its future success.

Many aspects of a firm's culture will be apparent in how the firm treats its people — and so this chapter on people will help firms determine their underlying values and thus the nature of their culture in this very important regard. Hopefully, your mission statement (see chapter 2) will have at least some indication of how you regard your staff.

Vive la différence — celebrate people's differences

One of the main problems facing property businesses is the inability to recognise and value the different styles and approaches of people. When you are used to tackling things in a particular way, it is hard to accept that someone else might achieve the same result by doing something fundamentally different.

Partnership discussions and all people-related matters become much easier to deal with when you recognise and celebrate these differences rather than fighting against them. The core skill of empathy (seeing things from the other person's point of view — see chapter 6) is vital in marketing and selling and is just as valuable in just about every other situation where you are trying to deal with people.

We do not have the time here to look at the wealth of psychological research that show different ways of thinking about people so I have selected just a ideas that I have found useful in property partnership environments. But please understand that I am not suggesting that people should try to change themselves or their personalities — but if you understand the basis of other people's thoughts and actions, then it is easier to modify your own behaviour to smooth communication between you.

My favourite (simple) model is described below — it suggests that people are essentially biased to a particular style:

- Dependent — These people are recognised by their smile. They like interacting with other people and need to be liked and appreciated by other people. It is important to dependent people that they feel they have a friendship with others.

- Detached — Unlike dependent types, detached types often have a neutral expression and tend not to smile or indulge in small talk. Sometimes they are considered cool or aloof. These people are not concerned with (and may find it uncomfortable) to deal with the social aspects on interaction — they prefer to focus on the task at hand, the facts and the figures and can be seen as being curt and unfriendly, instead of concise and efficient.

- Dominant — Sometimes these people are recognised by their frown. They will often be the ones who control the conversation and do most of the speaking. They like to believe that they know all the answers, are in control and are calling the shots.

These "thumbnails" of the different types are very general. However, they illustrate the potential difficulties when different types try to interact. Imagine the poor dependent person feeling unloved and unappreciated by an efficient detached person or overpowered by the dominant person. Or imagine the detached person squirming in their seat as a dependent person chatters away about their family and friends. By recognising the different types and modifying our own expectations and style accordingly, we can ease the communication process for all concerned.

Another model considers the different way in which people think and deal with the world around them. The various types — and the best way to deal with them — are shown below:

- Synthesists — Are curious, creative thinkers and like to speculate with "What if?" questions. It is best to listen to these people until they have shared their thoughts with you.

- Idealists — Have lofty goals and high standards and strive for agreement. You will succeed with these people if you link the greater goals of the firm and quality standards with whatever you are discussing.

- Pragmatists — Are flexible, resourceful and practical people, who seek an immediate pay off. Concentrate on short-term aims and results to get them on your side.

- Analysts — Thrive on accuracy and attention to detail and like to gather data and adopt a methodical approach. They may be uncomfortable pursuing actions based on "gut feel" decisions. Make sure you have a logical plan, based on objective information, and avoid any errors or gaps in your analysis.

- Realists — These are fast-moving doers, with a short attention span. They rely on their senses — sight, sound, taste, smell and touch. Keep things really short with these people — short summaries of information.

Some firms have adopted more sophisticated methods of understanding the different personalities and styles of their partners and people by using psychometric tests. A common approach is the Myers-Brigg assessment, which provides people with a four-letter summary — so, for example, an ENTP person exhibits the following characteristics:

- Introvert–Extrovert
- Sensory–Intuitive
- Thinking–Feeling
- Judging–Perceiving

There is a great deal of information online about the various types and how best to adapt your style to make communication with other types more effective.

Checklist and actions

- How well do you know your partners and staff? Have you explored assessments to see where and how you differ?

- Are there ongoing difficulties in communications among particular individuals? Might these be eased by analysing the different types and styles of the individuals involved?

- Does your firm (or the management team) comprise too many of the same types of people?

- How can you gain the input of people with different styles into your business?

Table 8.1: Dealing with "difficult" behaviours

Type	What do they do?	Why are they like this?	Potential strategies
Ogre (Shrek)	• Hostile–Aggressive • Shout and insult • Overwhelm by evoking buried childhood fears	• Rapid/confident decision makers (realist thinkers) • Need to feel powerful • Vein of (unconscious) self-doubt	• Simultaneously show own strength but show you are not a threat. • Watch your posture and projection. • Humanise — use their name. • Do not argue — disagree. • Use interruptions. • If you cannot talk, write (informally). • Keep them posted. • Acknowledge their strengths.
Fire-eater (Volcano)	• Irritable and moody • Hot tempered and explosive • Sudden rages	• Feeling personally threatened while at the same time under pressure to take some action	• Leave them to calm down. • Neutral statements to stop the tirade. • Repair the threat (use terms that reinforce their seniority). • Break off the interaction — allow them to be alone for a short while. • Understand mood patterns and triggers. • Help them understand the effects of their outbursts. • Humour: "Glad I got you on a good day".
Expert know it all (Bulldozer)	• Closed to other ideas • Lecturing mode • Superior • Acts like a teacher • Always right • Aggressive • Unforgiving	• They are very good at their job and are intolerant of anyone who does not reach their high standards	• View them as a source of great knowledge and an opportunity to grow. • Show respect for their knowledge. • Prepare and methodically plan as well as you can. • Provide as much concrete analysis as you can. • Select tasks that are of little interest to him/her and with clearly defined responsibilities that can be measured. • Use questions to point out problems — "explain it to me" and "describe how it might work in other areas". • Help them save face if things go wrong. • Cost/benefit analysis — You may have to give up!
Artful dodger (Invisible man)	• Stallers • Wafflers • Super-delegators	• Co-dependents (want everyone and everything to be ok).	Stallers • Ask their thoughts on how to improve. • Be positive about any negative feedback you receive.

Type	What do they do?	Why are they like this?	Potential strategies
	• Abdicate leadership • Avoid confrontation • Avoid anyone feeling bad • Fail to explain their expectations • "Dump" on you • Too many sidelines taking their time and energy • No supervision		• Don't reveal you are upset at feedback. • Ask them to make conflicts explicit. • Act like a problem-solving consultant. • Give verbal support to the right behaviours (agree). • Emphasise quality and service, not your own advancement. • Take decisions for them: "Here's what I plan to do". Wafflers • Watch for signs that you are pushing too hard. • Be personal and ask them about themselves. • Don't fight compromise — find a win–win. • Incremental small steps forward. Superdelegators • Let them know how you want to be supervised. • Negotiate your level of authority. • Coax out guidelines. • Schedule regular short meetings. • Agree turnaround times.
Powerclutchers, Paranoids and perfectionists (Ceasar and Marvin the paranoid android)	• Too much supervision • Detailed instructions • Check everything you do	• Need to be certain • Lack of confidence and trust • Irrational search for perfection • Over strong wish to be in charge • Fear of failure	• Uncover any hidden doubts they have about your abilities or trustworthiness. • Acknowledge their concern and show how you plan to avoid it in future. • Communicate in his/her style (see thinking styles). • Build trust by accepting fears and suspicions. • Welcome frequent check ups. • Probe for clues to fear points. • Emphasise contingency planning • Verbal support for risky decisions. • Try subtle teaching.
Scallywags, schemers and skunks (Devil)	• Scallywags — prey on others for own self-gain • Schemers — focus on own gain without considering the firm • Skunks — knowingly upset others	• Unscrupulous or offensive • Aberrant set of values • Fearful of others (need to strike first) • Inability to feel for others (sociopaths)	• Assess the situation — distinguish bad guys from the good guys and simpletons (honesty may show good guys are unaware of the offence they cause). • Clarify what you are up against — illegal or unethical, others feel the same, support for behaviour from above, your own goals, costs if you take action.

Type	What do they do?	Why are they like this?	Potential strategies
Hurt friends (Angry child)	• Crossed expectations • Behaviour blindness • Interaction accidents	• Failed to communicate properly • Misunderstanding • Unaware their behaviour caused upset • Reacted badly to something negative you said • Hurt feelings	• Disengage, if practical (resign — with a record of your achievements in tact). • Protect yourself — resist, express your feelings, consider counselling, document everything. • Blow the whistle — attempt collective security, attack the problem not the person. Expectations • Confirm your understanding of what you are expected to do/responsibilities in an informal note. • Recommend rather than assert. • Confirm with them before proceeding. Behaviour • Make a formal appointment to give unpleasant feedback. • Show that you recognise ambiguity — your interpretation may not be the correct one. • Help them save face — focus on what needs to be changed. • Describe the difficult behaviour (be specific and descriptive). • Restate assumption that they are unaware, watch for acknowledgement and provide support. Interaction • Assess — abrupt change for the worse, precipitating event, good relationship with others, own emotions are extreme. • Make a short notice appointment late in the afternoon, set the stage, comment on state of relationship, prepare to be dumped on, convey understanding without excuse or apology, state your intentions, move to problem solving (focus on the future).

Source: Robert Bramson (adapted)

Tantrums and time wasters — dealing with difficult people

Read the section above on "Vive la différence" first. The person might not be difficult at all — just different. The second thing to recognise that by "labelling" someone as "difficult" will make it harder for you to understand what drives that person's behaviour and will make you blind to the more positive behaviours that they exhibit.

People are rarely difficult deliberately. Usually there is some positive intent behind whatever they do. Their difficult behaviour can be the result of a number of things — some of which are negative and some are positive. Some aspects of their behaviour they are aware of and can control, and other aspects are the result of their personality or of things of which they are unaware.

Based on the book *Coping with difficult bosses* by Robert Bramson, I have developed a methodology that considers the various types of "difficult" people, considers why they are difficult and suggests ways in which you might better deal with their particular type of "difficult" behaviour. A summarised version is shown in the table 8.1.

Sometimes bullying can be a problem. It helps if you recognise the psychological rewards of bullying. By demonstrating how weak others are, the bully feels more powerful, less weak and out of control and buttresses their own strength and power. Bullies are effective because they cause others to react in a specific way. It is important that management recognise and deal with bullies — otherwise they may find themselves in an Employment Tribunal facing a hefty fine and negative publicity.

Therefore, adopt a more thoughtful approach to dealing with difficult people. And avoid RASP — Reciprocal Attack Spiralling Phenomenon, which will result in a very unhappy ship.

Time to talk? Internal communications

The property industry is intrinsically a people business. While expertise, reputation and profile are important, property transactions start with relationships. By and large, the property industry is also a pleasant business — while it remains very commercial, most people are polite and respectful to one another. Staff are generally treated well but, unfortunately, staff are not telepathic. And one of the main faults in managing people in property businesses is a lack of communication with staff.

Partners sometimes forget that while they have seen all the numbers, done all the thinking and feel confident about the future — they need to bring the troops along with them. Otherwise, staff can end up feeling left out and ignored. And staff who feel left out and ignored are likely, at best, to be poor producers and, at worst, to walk out the door — sometimes taking your clients with them.

While many people think that an active and planned approach to staff communications is the preserve of the large firms, it is actually a topic that requires attention for relatively small firms. And especially so if they are spread across a number of different locations or organised into self-contained departments.

At the most basic level, you need to ensure that everyone is aware of the aims of the business, its general strategy and their role in its delivery. An annual staff meeting is a good starting point. You also need to repeat the message regularly — perhaps quarterly.

This should also be reinforced with a short monthly email reassuring everyone that things are on track. The A* management teams will even make the effort to thank the troops occasionally and give particularly good effort or results a special mention in despatches. So that is the formal top-down communication process, but what about hearing — from the grass roots — what they think?

Contrary to common belief, it *is* unrealistic to expect young surveyors and support staff to raise their hands and ask questions of the partners in an all firm meeting. Therefore, to learn what staff really think you need to find less intimidating environments. Larger firms might invest in staff surveys to gain feedback, but face-to-face communication is always the best policy. Some firms have excellent programmes, such as senior partner suppers and team awaydays to do this.

You also need to ensure that there are regular short meetings for particular departments, teams and offices. This way, the head of department (providing they have some basic communication abilities can "cascade" down the thinking of the senior management team. To ensure consistency here, you might offer them a few bullet points of the key points to convey. You should also encourage the teams to ask questions and to offer their views.

Inevitably, there will be some department heads who are brilliant at briefing their team and keeping everyone involved and motivated. The reverse is also the case. So do not expect all team leaders to have the same people and communication skills — find another way for their team to be kept in the loop though.

On occasion, make sure that staff are allowed to mix with and hear from other departments and offices. Otherwise, you will create a firm of silos and your chances of your staff being great ambassadors for all aspects of your firm and facilitating cross-selling will be reduced.

There are many other ways for you to communicate with your staff apart from time-intensive meetings. Even fairly small firms have tried internal staff magazines. Most will have some form of intranet, and everyone uses email. Some managing partners write a short (two or three lines) email each week communicating some particularly interesting or useful piece of news about what is going on. These are all tools that can be used effectively to promote good internal communications.

Do not forget the most effective internal communications tool of all — MBWA (Management By Walking About). There is no substitute for having people taking the dramatic step of leaving their offices and walking round among their staff, saying "Good morning" and showing an interest in what their staff are doing.

Checklist and actions

- Do you tell all members of the firm about the firm's aims and strategies?

- Do you regularly update everyone on the firm's progress and key developments?

- Are there regular team meetings in all departments and offices?

- Are comments and questions from the grass roots communicated up to management?

- Are there opportunities for everyone to learn about what is happening in other departments and offices?

- Are tools, such as internal newsletters, the intranet and email, used to communicate news?

- Do partners make an effort to walk around and talk informally to their staff?

Talent spotting — recruitment

Recruitment is an expensive process. Apart from the recruitment agency fees, there is a huge amount of time invested in interviewing and training new staff. So before you look at how to improve your recruitment, you should look at the rest of this chapter to see what else you can do to improve the retention of your existing staff!

Growth is impossible if you cannot recruit the people you need when you need them. While a firm can cope for a while with insufficient numbers of the right calibre people, in time the quality of work and client service will suffer and your existing people may become overstretched and decide to leave. Therefore, recruitment is fundamental to support your firm's growth plans.

A strategic (in support of the business plan) and proactive approach to recruitment is essential in all surveying companies. There are three main types of recruitment that you need to accommodate:

- graduate recruitment;
- qualified staff (professional and support); and
- senior or lateral hires.

It may be that you need to use different agencies (both recruitment and headhunters) depending on the task at hand. You may also decide that you need to place advertisements in local or specialist property media. It is important to involve your marketing specialists in this process as clients and potential clients may also see these ads so they must reflect the firm's identity and convey consistent messages about the firm. In addition, you may need help on identifying the most appropriate media and designing an advert that is effective.

Whoever is involved in recruitment should have some training or guidance on how to interview — to explore whether people have the required competencies and experience, to ask appropriate questions and to ensure that employment and discrimination regulations are observed. Detailed records must be kept (that comply with the Data Protection Act) in a secure place so that you can a) track the effectiveness of your recruitment programmes b) respond to any later questions about the recruitment process and c) refine your future recruitment programmes. Some firms find it helpful to use psychometric tests and/ or assessment centres to help improve the effectiveness of their recruitment, however, these can prove expensive.

Remember too that recruitment is a matching process, as there are two essential things that you are trying to assess during recruitment:

a) Whether the individual is suitable for the role (their technical skills, competencies, experience, track record, management ability, personality etc) — they are selling themselves to you.
b) Whether the role and firm (culture) is right for the individual — you are selling the firm and the role to them.

Graduate recruitment

Graduate recruitment needs to be tackled as part of the overall graduate intake programme where you will need to provide the necessary training and development support over the two-year training contract. Most firms have a rolling annual programme of activities to ensure that all the key dates are met and that the firm is competing well with other firms seeking the same types of candidates. Some firms also assign mentors or buddies to new graduates to help them find their feet in the firm and as a source of advice during their training contract.

You may also need help from your marketing specialists in providing the necessary support to your HR people and the partners (and younger surveyors) charged with seeking out the best graduates. This could involve exhibitions (the milk rounds), talks at local universities or at group assessments or interviews. Some firms arrange open days where undergraduates come to the offices to have a look around and to chat to existing graduates to learn more about the firm.

Most firms will have a separate section of their website that contains information relevant to potential applicants (for example, profiles of past and present graduate trainees, criteria for candidates, schedule for applications, a blog by existing trainees) as well as an automated application form. While this area of your website needs to fit in with your firm's overall identity, it does need to be sufficiently different to appeal to the rather different needs of the younger generation who are much more computer literate and sophisticated in their expectations about a firm's use of technology. Some firms have recruitment programmes in social networks and virtual worlds such as Secondlife.

Selecting and interviewing graduates can take a lot of time. Therefore, it is important that you do everything possible to ensure that the process is as effective and as efficient as possible. The following areas should be addressed:

• Target universities — it may be easier to focus your efforts on a limited number of colleges where you are most likely to obtain the candidates that you seek.

- Candidate specification — the type of candidates sought.
- Job specification — outlining the nature of the work undertaken, including the options for working with different specialisms.
- Timetable — dates for applications, interviews and offers.
- Team — who has overall responsibility for managing the graduate intake in addition to those partners and staff who will be involved in the interviewing and selection?
- Training/briefing — to ensure that everyone on your team interviews to the same standards, explores the same issues with candidates, conveys the correct information and observes the numerous legislative regulations.
- Q&A — a list of standard questions and answers to help your team provide accurate and consistent answers to typical candidate questions.

You should keep detailed information about what time and cash is invested in the graduate programme. There should also be statistics about how many people apply and from what universities and how many interviews convert to offers and how many offers are accepted. You should also track how well gradates from different universities and backgrounds get on once they have joined the firm. All of this information will help you refine and improve your graduate recruitment programmes.

Qualified staff

The need for qualified staff will vary depending on the turnover rates of staff and on how the business grows and develops. Sometimes you will need to recruit qualified staff to replace those who are leaving, sometimes it will be to add resources to a fast-growing team and sometimes it will be to acquire expertise or experience that the firm currently lacks.

It will be difficult to obtain the right quality of people if you are in a big rush, so plan ahead for the staff you are likely to require and talk through your future requirements with a trusted recruitment agency. You may also find that having such a relationship with an agency means that you can negotiate favourable rates and a preferential service. Remember too that in your eagerness to recruit a particularly good candidate, you should not forget about the firm's policies and the salary and promotion criteria operating — otherwise your shiny new recruit may cause resentment and upset among your loyal existing staff.

Depending on the nature of the staff you are recruiting, you should always have a clear job description and a detailed person specification. The salary and benefits package must be clear (or limits set if there is room for negotiation) and so must the contract of employment.

You should decide, in advance, who will be responsible for managing the recruitment process and who will be involved both in interviewing and in selection. Where you are recruiting people with skills that are not presently at your firm, you will need to find a way to assess those skills during interviews, which may require help from a recruitment specialist and/or someone outside the firm who is familiar with the area. Sometimes, firms need to make a special effort with prioritising recruitment activities, as many a good candidate has been lost as a result of slow decision-making or convoluted interviewing procedures at firms.

Senior or lateral hires

There may be occasions where you need to recruit at a senior level — perhaps a partner to replace a retiring person (what happened to your succession planning?), to head up a new team or to establish a presence in a new market. Sometimes, you have to assess senior people who are joining the firm as part of a merger or acquisition (see chapter 9 for more information about mergers). These assignments are particularly tricky as there is much at stake. If it goes wrong you may cause damage to the firm's reputation, to the stability of existing teams or to critical expansion plans.

A critical issue will be the relationship between the potential senior hire and their existing clients and whether they have restricted covenants preventing their following from moving with them. Similarly, the supervision, management and business development skills of a senior hire are likely to be a major concern.

Negotiating the financial and partnership package will also be a challenge, particularly if the individual requires incentives that are beyond or different to those that the firm currently provides. Having clear expectations — that are articulated and discussed during the recruitment process — will minimise the chances of mismatches or surprises for both parties. In addition, the negotiation process may be protracted and take up a lot of management time.

Checklist and actions

- Does your business plan (and departmental or office plans) show recruitment requirements for the year ahead?

- Are there detailed statistics on the costs, time and success of recruitment processes?

- Is there an agreed graduate recruitment plan?

- Is responsibility for recruitment, interviewing and selection allocated?

- Are there detailed person and job specifications for all recruitment programmes and detailed information on the screening, interviewing and appointment processes?

- Are you getting the right service and support from recruitment agencies and other external advisers?

- Are advertisements (and websites) produced in line with the firm's identity but tailored to the particular vacancies and media?

Keeping the saws sharp — training and development

With staff taking up to 50% of a property business's costs and with recruitment being such an expensive activity, it is vital that your firm has a planned and structured approach to training and development. Training and development is an area that is considered carefully by those about to join your firm, it will be a major source of satisfaction (or dissatisfaction if it is wrong) with your existing staff and an important tool in helping your firm grow the skills to serve clients and support the firm in the future.

A starting point is to do an analysis of the skill areas where there is the most need — ie, those that are holding back the firm (and the person). This enables training to be applied in a targeted way, rather than like a sheep-dip. It is important to recognise that different people learn in different ways. A simple way of thinking about this is the learning styles model — and it is relatively easy to assess which of the various styles is dominant in each individual. The following model was developed by Peter Honey, who explained that the learning cycle (how people learn from experience) comprised a number of stages: do, review, conclude and plan:

- Activists
  These people are "doers" and learn best from the something that happens to them (passive) or an experience that they deliberately seek out (active). Training for these people needs to allow them to "have a go" at something, become involved in short activities such as role plays, where they are thrown in at the deep end and where there is plenty of excitement and a range of changing things to tackle, usually involving people.

- Reflectors
  These "reviewers" take a non-judgmental look at what happened in the learning experience. They learn best when they can stand back from events and listen and observe, where they can carry out research or analysis, where they can decide in their own time and have a chance to think before acting and where there is an opportunity to review what they have learned.

- Theorists
  These people draw conclusions from the thoughts and notes made at the review stage, to identify lessons learned. Theorists will learn best when they are intellectually stretched, where they can question assumptions or logic, where the situation has a structure and a clear purpose, where they can deal with logical, rational arguments with time to explore them and where they are offered interesting concepts, although they might not be immediately relevant.

- Pragmatists
  "Planners" will plan and test the lessons learned from the conclusions so that they can be related and applied to similar situations in the future. They will learn best when the techniques have obvious practical benefits, where they can implement what you have learned immediately, where they can try out and practise techniques and where they see an obvious link between the subject matter and a real problem or opportunity.

While many firms still focus on training that is classroom based (often provided by an external provider at an outside venue) or delivered by senior members of the firm, there are many other ways to ensure skills remain sharp. As you will have gathered from above, "lecturing" people will not be the most effective style of learning for many people. A study by Ebbinghaus found that 90% of what was learned in a class was forgotten within 30 days and 60% forgotten after one hour! So it is best to find training methods that gain greater involvement from those who are learning.

Developments in online training (e-learning) mean that standard programmes can be delivered to large numbers of people at different locations easily and cheaply. There are also web seminars (webinars) where you can use PCs and telephones to enable people at disparate locations to hear what a speaker in a central location has to say while watching Powerpoint slides and other visual materials on their PC. In many firms, much training is done "on the job", with people providing specific help to individuals — this can be made much more effective if standard processes and explanations are documented and made available on the firm's intranet.

Some firms find that quality accreditation will enable them to develop and run training programmes in a structured way that is focused on the needs of the business and incorporate "best practice". A good example of this is Investors in People, which requires every member of the firm to have a training and development plan that links their role to the firm's overall business objectives.

Many firms appoint particular members of the secretarial and support teams, and particular members of the professional teams, to act as training co-ordinators, which helps spread the load and ensures that the firm remains aware of emerging training needs.

Training needs analysis might be undertaken at the outset in order to assess the views of all members of staff regarding their roles. This can be done through a questionnaire but is probably best done by having representative groups of staff gathered together for a short while for an informal discussion (such as a focus group).

There are various elements of a training programme in a property firm that require attention.

Induction training

You should have agreed standard induction programmes for all new staff when they join. Obviously, these programmes need to be tailored to the particular types of new joiners. As such, the induction programme for new graduates will be substantially different from that of new secretaries. However, the core material in all induction programmes will cover:

- About the firm — its history, values, organisation, work, services and clients.
- Technology — standard and specific systems for office support, technical work, billing etc.
- Procedures and policies — advice on observing the relevant quality standards, health and safety procedures, environmental policies and so on.

Many induction programmes go beyond this and include things such as a tour of the offices, introductions to key staff, presentations from all departments to ensure that every new joiner is aware of the key personnel and major services and clients in each department.

Remember that if you cram all the induction training into a new joiner's first few days they are likely to be overwhelmed so it may be more sensible to have a series of short (eg, half-day) sessions spread over the first few weeks. Similarly, it may be sensible to schedule in follow-up sessions to check that new joiners have all the information they need and to determine how the induction process can be improved.

Technology training

While every member of your firm will be using technology, their levels of competence may vary significantly. It might be worth undertaking a technology skills audit of all professional and support staff to ensure that everyone is working as efficiently as possible. Some firms offer a regular programme of technology training sessions (sometimes across lunchtimes) to ensure that people stay up to date with both standard systems and those specific to your firm and their practice area. If there is sufficient need at your firm it may be more efficient to invest in e-learning solutions so that staff can learn at their desks and at a pace that suits them. Older and more senior members of the firm may benefit from desk-side training, where a member of the IT team spends a short while helping them get to grips with new systems in a modular fashion and perhaps targeting those areas that are of most value.

Many firms fail to realise the efficiency gains of investment in new technology systems because they do not provide sufficient cash and time to the training programme. When new systems are installed, there will need to be a major programme of training (and extra support) before the new system is installed and as the firm transitions to the new system. These extensive training programmes (and the necessary documentation and/or help posted onto your intranet) should be part of the overall project plan and your suppliers should be involved in the training programme.

Professional/technical training

New graduates will require a detailed training programme of both externally sourced and internally provided training to help them prepare for their professional examinations. Qualified staff will need support in meeting the requirements of their professional bodies Continuing Professional Education (CPE). Other members of your firm may also belong to professional bodies that require ongoing CPE.

While most firms look to their staff to maintain their own CPE records, they will often provide some support with selecting the appropriate courses. Where there are major changes in professional practice, it may be necessary to arrange for large numbers of staff to be trained in new processes or standards.

Compliance training

Usually, compliance training is tackled as part of the induction process but sometimes — for example when the recent money laundering regulations were introduced — you may need to ensure that all professional staff are trained in new procedures. Your compliance officer should be aware of what training requirements are and there are often cost-effective, online training programmes available. If your firm has a quality programme, you will need to provide training in this on a regular basis.

"Soft" skills training

Helping people to become more effective in their soft skills is becoming increasingly important. The types of soft skills are hugely varied but common requirements here might be: communication, interviewing, supervising staff, presentations, networking, business development, dealing with the media, dealing with difficult clients and negotiation. There is plenty of material about such skills in chapter 6 on selling. Often, you will need to obtain external advisers to provide training in these subjects and you will often need to tailor the material to address the particular challenges and situations most commonly met in your firm.

These types of programmes can be delivered online but with soft skills training it is particularly important to allow face-to-face contact for questions, for individual issues to be tackled and to allow staff to "practice" their new skills in role plays in a safe environment.

Management development

As markets change and firms grow larger, there is an increasing need for structured management development programmes. The senior people running firms are unlikely to have had any formal training early in their career and have managed to grow the business — while it was small — relying on common sense, commercial nous and a keen eye for the numbers.

Often, firms will reach a plateau where nothing they seem to do helps them grow any further. Or they hit a number of crisis — such as the inability to reach the different types of clients that they want, or to make new offices and teams work, or finding that senior staff continue to leave just before they are invited to join the partnership or that the practice can no longer do the work it has always done and still make a profit. It is not only those who are managing the whole firm who find that they lack the necessary management skills, but also people managing offices or even quite small teams can find that they are inadequately equipped to deal with the management tasks that are beyond their technical professional skills.

There are many ways to tackle management development. There is, of course, the quick fix approaches. This means sending off key people for a few days on an external course. The trouble is, if the training is general and not tailored to the special needs of a partnership or the property sector, then it is unlikely to be accepted by your partners. Alternatively, you can hire a management consultant to tackle the specific issue and hope that some of his or her skills are magically absorbed into your people. That is a possibility, but some consultants are more concerned with keeping their clients in the dark so that they are guaranteed their next assignment.

You might encourage your entire management team to go through a structured learning programme that tackles the main elements of management — strategic analysis, financial management, human resources management, knowledge and information management, technology and marketing — over a series of months. While it may seem an expensive route (in both cash and time) consider the cost of continuing to run your practice in an unprofessional way! Coaching (see below) is also a particularly effective method of helping people to take responsibility for learning and development throughout critical times in their careers (eg, after a promotion, as they take control of a team or on admission to partnership) or to help fast-track individuals through a range of new learning challenges.

Training aims to help people learn new knowledge, develop new skills and subsequently work in a different way. This means that training is unlikely to have an immediate impact as it will take time for people to recognise that they need to change, accept the desired changes and build up the confidence to attempt to use and practice their new skills. This process is conveyed in figure 8.1. What it means is that you must be patient with people when you require them to learn and do things a different way.

Figure 8.1: The competence curve

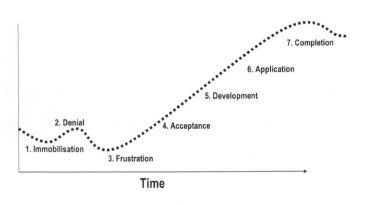

There is more about change management — for both individuals and organisations — in chapter 9. Whatever training and development programme you have at your firm, it is important that you have some way of assessing the effectiveness. After all, training is a costly business — there is the actual cost of the training and the significant amount of the time that your people spend in training. Of course, after each learning activity you can gain immediate feedback through a simple satisfaction form or, even better, a few weeks after the training you can ask the learners what they recall from the training and what they have done since to implement the new ideas and what results they have achieved. Some firms require each person attending a training session to provide a short internal presentation of the key points to their colleagues on their return.

Checklist and actions

- Who is responsible for training at your firm?

- Is there an annual plan and a budget for training?

- Have you undertaken a training needs analysis to identify the priority areas for the business in the next year?

- Does your appraisal system (see below) provide input to your training plans?

- Do you measure the effectiveness of your training?

- Do you have training programmes to deal with the different areas of:
    - Induction? (a firm's systems and procedures).
    - Technical and professional skills?
    - Commercial and client skills?
    - Interpersonal skills?
    - Compliance?
    - Technology?
    - Management?

Encouragement through coaching and mentoring

Now some of you may think that coaching and mentoring is rather too "woolly" to be taken seriously and that you just have to throw the odd encouraging remark at people in order to tick the box on being an effective coach or mentor. Sorry, you need to think again.

While there are many cowboys out there who purport to be professional coach/mentors while convincing you to part with large sums of money, coaching is increasingly becoming a profession in its own right — drawing on models in education, psychology and, of course, sports motivation.

Professional coaching is about achieving significant change and measurable improvement in performance within a specified period of time. It is *not* to be confused with counselling, which is more about personal remedial action.

The official definition of coaching is: "A process that helps and supports people manage their own learning in order to maximise their potential, develop their skills, improve their performance and become the person they want to be" (Parsloe). The official definition of mentoring is: "Off line help by one person to another in making significant transitions in knowledge, work or thinking" (European Mentoring Center).

There is a great deal of research supporting the value of coaching and mentoring. For example:

- 50% of managers found coaching and mentoring more effective than traditional training courses.
- US research showed, on average, that only 10–20% learning through training transfers into people's work.
- The Government's Declaration of Learning identified 14 individual and business benefits of coaching.
- A CIPD survey of 800 UK training managers revealed that 87% now use coaching and mentoring.
- Learning on training courses in the workplace is now widely recognised as only one of at least 50 ways to learn.
- Coaching and mentoring is now the third most frequently used tool in the corporate armoury, after on-the-job training and training courses.

In essence, coaching and mentoring provides an experienced and trained independent person with whom to discuss work (both in the short and long term), identify and discuss development options, obtain tools to undertake self-assessment and improve performance in a confidential environment.

Being one-to-one, it is totally tailored to address the needs of the specific "learner" — whether they require help with a specific situation, an objective sounding board during a time of transition, to gain ideas on challenging issues, to maintain a high level of performance, to clarify different options or to develop the ability to manage future learning in a structured way. Although it is focused on the individual needs, it invariably leads to improved performance in the current role, which is why organisations benefit too.

There are many benefits of coaching for individuals, for example:

- Improved performance.
- An unbiased and confidential sounding board on workplace issues and situations.
- Increased awareness of learning.
- Objective assessment of skills and potential.
- Maximised learning opportunities.
- Creation of a Personal Development Plan (PDP).
- Assistance with actions from appraisals.
- More positive feel about development.
- Increased morale.
- Positive impact on work and private life.
- Increased motivation.
- Better equipped to accommodate changes in the workplace and role.
- Significant business benefits, such as better team leadership, improved business development skills, better management abilities.

Typically, formal coaching will comprise a series of regular (perhaps monthly) sessions. There is usually a preliminary meeting — lasting around two hours — where the coach/mentoring process is explained, goals are set, questions are answered and initial ideas on what to tackle are discussed. There may be self-assessment exercises. A simple model to guide coaching discussions is offered by John Whitmore:

- Establish **Goal**
- Check **Reality**
- Explore **Options**
- Confirm **Will** to act

Leadership was covered in chapter 3 but a nurturing leader actively promotes the growth of group members in terms of skills, knowledge and emotional well-being. Supportive communication is a communication style that delivers the message accurately and supports or enhances the relationship between two parties. He or she consistently attempts to make people feel good about themselves rather than cut them down with insults and hostile humour. Remember that a mixed message comes about when you say one thing but your tone and body language say something else!

Whereas coaches are usually external experts, mentors are more likely to be drawn from those within your firm but who do not have line responsibility for the individual(s) they are mentoring. A mentor is an individual with advanced experience and knowledge who is committed to giving support and career advice to a less experienced person. Much of the careers advice centres on upward mobility. A mentor is a helper and confidante, as opposed to a rival or detached manager. A protégé often does not report to his or her manager in a hierarchy. Mentors help explore, in a non-threatening way, the ideal versus the reality so that the gap can be examined. Through careful questioning and non-judgmental listening, they can explore issues and help set the bar just a little higher each time.

In addition, the mentoring process is more likely to be less formal and more sporadic than coaching, as there are unlikely to be specific goals. In property businesses, often new staff will be allocated a more senior and/or experienced professional to talk to and new partners will be allocated to a more mature partner to act as an objective sounding board and to discuss performance (and partnership politics) over the occasional glass of beer or wine.

Remember that great leaders also act as mentors — but they need the following attributes to succeed:

- risk taker;
- trustworthy;
- knowledgeable;
- influential;
- committed;
- emotionally stable.

Checklist and actions

- Does your business plan (and departmental or office plans) show recruitment requirements for the year ahead?

- Are there detailed statistics on the costs, time and success of recruitment processes?

- Is there an agreed graduate recruitment plan?

- Is responsibility for recruitment, interviewing and selection allocated?

- Are there detailed person and job specifications for all recruitment programmes and detailed information on the screening, interviewing and appointment processes?

- Are you getting the right service and support from recruitment agencies and other external advisers?

- Are advertisements (and websites) produced in line with the firm's identity but tailored to the particular vacancies and media?

- Do you use coaches and mentors?

What did I do wrong? Appraisals and feedback

"Performance management is establishing a framework in which performance by individuals can be directed, monitored, motivated and rewarded and whereby links in the cycle can be audited." Mabey and Salaman

A vital part of learning (and motivation — see below) is regular feedback on how well you are doing. Too often, staff work extremely hard for years without having any idea whether they are considered to be good, bad or indifferent. A common complaint is that senior staff never say: "Well done" or "Thank you". In the best selling book *The One Minute Manager* the authors offer a brilliant technique called "one minute praisings" and "one-minute reprimands", which encourages supervisors to always immediately articulate when they people do the right things and when they do the wrong things. In a fantastic book called *Time to think* by Nancy Kline it reminds us that relationship research shows that we should balance negative feedback with positive feedback — and that the ideal ratio is five positive feedbacks to one negative! Interesting as most of the professions are trained to spot the mistake and only express any opinion when they "catch someone doing something wrong". So the challenge is to work harder at "catching people doing something right" and actually commenting on it.

Charles Seashore and Edith Seashore suggested that: "Feedback is information about past behaviour, delivered in the present, which might influence future behaviour". Positive reinforcement is increasing the probability that behaviour will be repeated by rewarding people for making the desired response.

Hence, everyone needs feedback on a constant basis. Many partners are surprised that their staff have not developed telepathic skills and when faced with an unexpected resignation remark "But we thought you were doing a fantastic job and that you had a great future with the firm" to the consternation of the resignee who had no idea how he or she was thought of by the partners. So, a critical lesson, make sure you give copious and regular feedback. And do not leave it to the annual appraisal — do you really want someone doing the wrong thing for a whole year?

The appraisal system is an important element of the firm's overall ability to manage the performance of its people. This is shown in figure 8.2.

Figure 8.2: Stages of a typical performance management system

DEFINITION OF
BUSINESS ROLE
• Job description
• Objectives of department/group

FORMAL ASSESSMENT
AND REWARD
• Annual assessment
• Link to pay

PLANNING PERFORMANCE
• Individual objectives
• Development plans

DELIVERING AND MONITORING
• Ongoing manager support
• Ongoing review

Many property professionals see annual appraisals as, at best, a boring paper exercise and, at worst, a complete waste of time. But the appraisal is not about you; it is about the person being appraised. It is one short hour each year when the focus is not on the firm, or the partners, or the clients but on the individual member of staff. Where they can have their supervisor's full and undivided attention to talk about how well (or not) they are performing, what they hope to achieve from the future and what they need to do to develop and/or progress further. So please devote your full attention to appraisals and show a little more grace in their execution.

Larger firms will no doubt have structured appraisal processes with detailed forms that consider competencies across a number of professional, technical and personal measures. But remember that the appraisal process — even if it is used to support salary reviews and bonus payments — is primarily about those being appraised. They are a vital tool in deciding what training and other development is necessary for each individual over the coming year — so a good appraisal process will make it much easy to set out your annual training plans (see above).

However, a structured appraisal process and forms is not always necessary. You can be just as effective simply by asking the appraisee, about a week before your meeting, to consider the following questions:

- What have I achieved over the past year? (Against any objectives set as well as other things.)
- What did I not achieve over the past year?
- Where are my strengths and weaknesses?
  - Technically?
  - Professionally?
  - Commercially?
  - With clients?
  - With colleagues?
  - With the firm's systems and procedures?
- How do I improve and develop in my weaker areas?
- What do I want to achieve over the next year?
- What do I need to do over the next year — and what support do I need — to achieve these goals?

The appraisal meeting should be driven by the needs of the appraisee although if there are particular issues or problems that you wish to raise you must ensure that these are included. While formal 360 degree feedback exercises might be beyond the resources (and patience) of your firm, it is a good idea to ask a few of your colleagues, the appraisee's peers and their subordinates for constructive feedback as well — it helps to get a rounder view of someone and specific examples should be used rather than sweeping generalisations. It will also help if, during the appraisal, you can award a grade for the appraisee in different areas — a simple scale, such as exceeding expectations, meets expectations, needs some development or needs a lot of development — is adequate. The purpose of these grades is to help the appraisee (and yourself) identify those areas on which to concentrate and to develop ideas on how to address them over the coming year. It will also enable them to compare their performance (and improvement) at the next appraisal meeting. Such grades can also be used in assessments for pay increases and promotion — providing you have a consistent approach across all staff or groups of comparable staff.

It is good practice to keep a note of what was discussed at the appraisal meeting and to ask the appraisee to review the notes and add any further comments of their own. A copy of the appraisal form or file note will be kept on the individual's personnel records and should therefore be checked for compliance with employment legislation and kept in a secure location.

As a summary of what you need to do at an appraisal:

Appraisal interview checklist
1. Purpose and rapport.
2. Factual review.
3. Appraisee views.
4. Appraiser views.
5. Problem solving.
6. Action and objective setting.
7. Document the key points and gain input/agreement.

Also, you need to provide feedback on a much more regular basis than annually. The following figure from *The One Minute Manager* is useful here.

Figure 8.3: The one-minute manager

Set goals; praise and reprimand behaviours; encourage people; speak the truth; laugh, work, enjoy

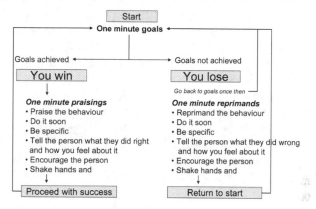

Source: Ken Blanchard and Spencer Johnson

Checklist and actions

- Are all those responsible for staff aware of and able to give regular feedback to their reportees?

- Do you have an appraisal process? Is everyone aware of and supportive of the process?

- Do you need to invest in some appraisal skills training?

- Are annual appraisals undertaken?

- How do staff feel about appraisals?

- Are partners appraised?

- How are appraisals linked with your reward, promotion and training systems?

Why should I? Motivation

Most people think that motivation is affected by money. Certainly those who are used to working with agents will have this impression. While money is a key motivator for many people, often it is not the only motivator. Many property professionals are motivated by a host of other things such as: how they are regarded by their peers, doing what they consider to be interesting and worthwhile work, delivering a high level of client satisfaction or contributing to the future success of the firm. So in order to understand how best to motivate people, we need to have a better understanding of the concept of motivation.

Motivation is a big and complex subject. We will start by reviewing some of the basic theory before considering its relevance to the property profession. A motive is a person's reason for doing something and comprises three elements:

- Direction (what a person is trying to do).
- Effort (how hard a person is trying).
- Persistence (how long a person continues trying).

When we consider a property partnership, the first priority is to have a clear strategy and plan so that people understand what the firm is trying to achieve and their particular role and contribution. The specific goals of the individual need to be considered too — if they are unclear, then it will be difficult for them to motivate themselves to pursue them. Effort and persistence will be affected by the recognition and reward systems (see below) — it may no longer be enough simply to say "If you want to be admitted to the partnership then you must develop new clients and new work". People need more structured goals, more tangible measures and more frequent feedback on their successes and failures.

Some people will be familiar with Herzberg's model of hygiene and health approaches to motivation. This means that you should ensure that the basics are satisfactory (eg, job security, pay and benefits, working conditions, supervision and autonomy) before trying to increase motivation through "health" issues, such as achievement, recognition, job interest, responsibility and advancement.

The culture of a firm is important for motivation too. Consider McGregor's "Theory X and theory Y". Do your partners adopt X or Y? Theory X suggests that people cannot be trusted. They must therefore be controlled and need financial incentives and threats of punishment.

If you apply this approach to intelligent and hard-working professionals, then they will often rebel against it. Theory Y suggests that people seek independence, self-development and creativity in their work. If treated correctly, they will strive for the good of their organisation — this is a better model to adopt when considering professional service firms. The later development of this theory — Social (Schein) — indicates that behaviour is influenced most fundamentally by social interactions. People are responsive to the expectations of those around them, often more so than financial incentives — and this suggests the need for cultural change and the development of team goals, plans and actions.

Intrinsic motivation is related to rewards based on the task — and this is often not appropriate in a professional services environment where extrinsic motivation (that is, unrelated to the task) is often a greater motivation — pride in a job done well, pleasure at seeing a client situation resolved speedily, satisfaction at achieving a good result in difficult circumstances and respect from fellow professionals.

We will now consider some of the other models of motivation. The expectancy theory (Vroom) attempts to explain how people choose which of several possible courses of action they will pursue. The choice process is seen as a cognitive, calculating appraisal of:

- Expectancy. (If I tried, would I be able to perform the task I am considering?)
- Instrumentality. (Would performing the action lead to identifiable outcomes?)
- Valence. (How much do I value these outcomes?)

Justice theories of motivation are concerned with equity theory, which suggests that a person is motivated to maintain the same balance between his or her contributions and rewards as that experienced by salient comparison persons. They will consider two types of justice:

- Distributive justice — whether people believe they have received or will receive fair rewards.
- Procedural justice — whether people believe that the procedures used in an organisation to allocate rewards are fair.

At the core here is the need for clarity around how people are rewarded for their efforts (how do you measure and reward them in your firm?) and to ensure that there are systems to ensure that all are treated equally.

Goal setting theories of motivation (Ed Locke) are also useful. "A goal is what an individual is trying to accomplish; it is the object or aim of an action". He offers some interesting insights:

- Difficult goals lead to higher performance than easy goals.
- Specific goals lead to higher performance than general "do your best" goals.
- Knowledge of results (feedback) is essential if the full performance benefits of setting difficult and specific goals are to be achieved.
- The beneficial effects of goal setting depend partly on a person's goal commitment (determination to try to achieve it).

But you should note that Earley suggested that goal setting may be harmful where a task is novel and where a considerable number of possible strategies are available to achieve it. This means that staff might need more support and structure in, for example, new business development activities with which they are unfamiliar.

In motivation, you might also consider the need cycle (you may be familiar with similar ideas in Maslow's hierarchy of needs). A need is an inner striving or urge to do something. It can be regarded as a biological or psychological requirement. Needs and motives function in about the same way because a motive is an inner drive that moves a person to do something. A need is one person's craving that drives the person to reduce the tension or satisfy the need by goal-seeking behaviour and this leads to satisfaction (a reduction of the drive and a satisfaction of the original need). Needs can be about:

1. achievement;
2. power;
3. affliliation;
4. recognition;
5. dominance;
6. order;
7. thrill seeking; and
8. security.

Day to day, motivation will depend on the psychological contract that people have with your firm (do they feel sufficiently involved in and committed to the firm that their additional efforts to develop new clients and work is an integral part of their role?) and the role models that they follow (do your senior partners demonstrate the marketing and business development behaviours you want more junior staff to emulate?).

As I said, motivation is a big subject. The theory can provide some helpful insights and tools, but each firm will need to examine a range of its cultural systems, beliefs and behaviour patterns before identifying how best to increase motivation towards achieving its aims.

Checklist and actions

- What is it that you want your partners and staff to be more motivated to do? Does this link with your business plan?

- Have you assessed — in a structured way — how motivated your staff and partners are?

- What appear to be the main issues affecting motivation and how are these different between the various groups of employees in your firm?

- How motivated do your partners and staff appear for technical, client-oriented, business development and management tasks?

- Does your reward system recognise and reward the behaviours that you are trying to encourage?

What's in it for me? Reward systems

Developing and adapting financial reward systems within a property practice is a complex science and delicate art. You need mechanisms in place to encourage all staff to achieve the firm's financial goals — whether this is measured in time billed, in deals completed, in invoices rendered or in monies collected. Of course, the danger is that the reward system focuses entirely on hard measures, such as time and fees, and neglects to encourage a wide variety of other behaviours that are vital to the success of a property practice, such as management, people development, client development and new business generation.

Partner reward

There is obviously a difference in the way that partners are rewarded compared with other staff, so we will look at this first. Historically, there was a tiered system in partnerships. Senior surveyors would be admitted to the partnership on the basis of being "salaried" for a while as they adjusted to their new role, demonstrated their effectiveness at generating work and thus earned their right to receive a share of the profits as an equity partner. Many partnerships used a lockstep system, whereby partners received an increasing amount of profit depending on their time in the partnership — and there were many advantages to this system (see figure 8.4).

Figure 8.4: Partner rewards — Lockstep

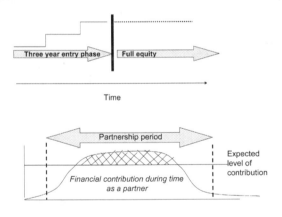

Source: Kim Tasso

A survey conducted by the Managing Partners Forum found the following:

An email survey was sent to 180 UK-based managing partners on 4 April 2002. 30 returned the survey by the due date (64% from law firms; 25% from accountancy; and 11% from property).

Findings:

- For 60% of firms equity partners represent between 7.5% and 12.5% of total staff, with 30% representing between 2.5% and 7.5%.
- For one-third of firms, partners are wholly rewarded via a points or units-based system for distributing the firm's profits. The majority favour a combination of profit sharing (80%), base salary (55%) or bonuses (20%).
- For those firms where partners are wholly rewarded via a points system, 80% use lockstep, with immediate past performance rewarded by additional unit allocations (40%) or cash bonuses (20%).
- The ratio of total reward between the most senior executive partners and a first-year partner varies from 1:1 (7%); 2:1 (35%); 3:1 (35%); or higher (23%).
- The spread in average level of total reward paid to partners last year was dramatic, with 30% of firms in the range of £125,000 to £175,000, but with no other range representing more than 10% of the firms surveyed.

However, there were a number of problems with this approach. It became difficult to deal with underperforming partners or those who were achieving way above expectation. It became difficult to reward partners who were undertaking valuable management work that did not have a direct impact on the fee income and profitability of the business. It became difficult to use the reward system to encourage partners to adopt different behaviours (for example, allocating work to others in the firm) that would enable the firm to grow. Then age discrimination legislation came into force, making many of the old assumptions about partnership and reward face legal challenge. As a result, many firms operate a different and more flexible system that allowed some of these things to be accommodated.

Rewards for non-partners

Developing reward systems for non-partners can be just as tricky. You will notice from the motivation section above that you need more than good rates of pay and a decent benefits package to keep people motivated. However, a poorly designed financial reward system is likely to cause dissatisfaction and can therefore have a powerful demotivating effect.

In addition to the financial reward, the system by which people apply for and achieve promotion should be transparent. This means that the criteria that are sought by the partners should be clearly articulated so that individuals can make a reasonable assessment of their ability to move forwards. It is important for motivation reasons — as well as employment legislation — that promotion decisions are made in a way that is fair and in a way that can be justified by reference to objective criteria.

Since 1999, the UK has had a National Minimum Wage (currently set at £5.35 an hour, with a development rate for younger people at £4.45). But pay is much more than the basic set rate for the work — there are variable and fixed elements, such as contributions to study, allowances for travel and supervisory or additional duties (such as being a First Aid officer), golden handshakes, relocation packages, food and accommodation subsidies, payment in kind (eg, as an alternative to a car), unsociable hours allowances, overtime, bonuses and incentives.

There are certain principles to apply in setting pay:

- A standard approach for the determination of pay across the organisation.
- As little subjective or arbitrary decision making as is feasible.
- Maximum communication and employee involvement in establishing pay determination mechanisms.
- Clarity in pay determination matters, so that everyone is clear about the rules and how they are applied.

You may have to undertake some form of job evaluation to ensure that you compare the relative value of different jobs in order to provide the basis for a rational pay structure. One of the major drivers of job evaluation over recent years has been the development of equal pay law to ensure that "like" work between male and female workers and young and older staff is considered fairly. Job evaluation often involved looking at different categories of work and developing grades within those categories: Common groups might be:

- Clerical and administrative.
- Manual and skilled manual.
- Technical and specialist.
- Professional.
- Executives:
    - middle management;

- general management; and
- senior management.

Typically, there will be various grades or bands within which pay is allocated for each category of job or work. On top of this, the firm will have a view about annual increases — from a basic "keep up with inflation" to a "reward for significant additional effort or results" and these payments are usually linked to the results of the annual appraisal. For example, you might agree that the annual minimum increase is set at 2% for inflation purposes but those who have worked well receive 3% and those who have excelled receiving 5%.

Ideally, your payment systems should be strategic, ie, they should follow on from and support the overall aims of the organisation. For example, if you agree that you want excellent staff in order to support excellent client service (and to charge premium rates as a result) then you will wish to have a pay strategy that shows you are prepared to pay a premium for staff. If, on the other hand, your culture and basis of client service is payment by results, then this will need to be reflected in your approach to pay as well.

Checklist and actions

- Does your strategic plan indicate the principles to apply in setting pay?

- Does your reward system support the achievement of the firm's overall objectives?

- Are the payment systems transparent and easily understood by staff?

- Do staff feel that the payment system is fair and equitable?

Rubbish! Dealing with underperformance

When staff are not working in the way that you require you need to take action. A common situation in a property practice is for everyone to know that a certain individual does not quite "cut the mustard" but no one is prepared to tackle the issue. This means that the individual may remain oblivious to the fact that they are not performing to the required level, those around the individual may start to feel resentment that they are "carrying" their colleague, clients may suffer as they receive a less than good service and the firm loses out as its investment in the individual is not paying off. The British condition means that often people will avoid confrontation in the vain hope that the situation will resolve itself. In reality, the longer the problem is left the worse it will become.

The first step must be to check whether the required performance criteria and desired behaviours and results are clearly articulated. These may be shown in the job description, the quality standards or the monthly financial results. Alerting the non-performing individual to written goals, job specifications, criteria or standards is a first step. Ask them to compare their performance against them. It becomes more difficult when the underperformance relates to attitudinal issues rather the measurable results.

Here is a 10-step approach to correcting errors:

Preparation
1. Get the facts — When? Which? How often? How important?
2. Never in front of others. Select an appropriate time/place.

Interview
3. Concentrate during the meeting on the impact of their errors.
4. Do not accuse or humiliate. Obtain their side of the story.
5. Look at the possible causes for the error — together.
6. Show them how to improve and offer/give help.
7. Set up an improvement action plan and fix a review date.

Follow up
8. Review specific performance since the interview.
9. Praise improvement. Discuss any continuing errors.
10. Set a new deadline for improvement and fix a new review date.

If it is a matter of disciplinary, then you will need specialist help, although I offer the following simple guide:

Discipline — the four Es:
*   Explain.
*   Establish the gap.
*   Explore reasons.
*   Eliminate the gap.

Sometimes the problem is caused by personality issues. For example, a manipulator attempts to influence people and circumstances to gain advantage through dishonest and unethical means. A personality quirk is a persistent peculiarity of behaviour that annoys or irritates other people. Then there are attention seekers, anti-control freaks, cynics about management and the passive-aggressive personality.

Have a look at the material above about dealing with difficult people, but call in an HR expert or a psychologist if it is a major issue.

Prompting an early, "interim" appraisal may provide a non-confrontational environment in which the individual can explore in a relatively safe environment what the expectations are and their perspective on their performance. Having a probationary period for new staff may support the effort to identify underperformance at an early stage. Sometimes there are extraneous factors affecting performance — the firm's resources, market conditions, changing technology or a difficult relationship with a supervisor, colleague or client. Sometimes it may be that training, coaching, greater supervision or mentoring support is required. On occasions, the individual may have a grievance and care must be taken that legally compliant procedures (which span fairness, representation and no more than three procedural steps) are followed.

When the situation does not improve with the appropriate support then it may be time to consider a formal disciplinary procedure. This must be written in the firm's "office manual" and available for all staff to see. It is at this stage that a good HR professional is required and easy access to employment lawyers who can advise on the appropriate procedures and remedies.

Checklist and actions

- Are expectations about what staff must do clearly articulated — in the firm's strategy, firm's policies and procedures, individual job descriptions, team and individual goals?

- Are there established procedures for dealing with employee grievances and employer dissatisfaction?

- Are there systems that allow you to objectively measure staff performance?

- Are there current (and legally compliant) processes to deal with underperformance and disciplinary matters?

Unsung heroes — support the support staff

Naturally, the focus of your attention will be on your professional staff as they are the most expensive resource and they directly affect the firm's ability to generate fees. However, no property practice could exist without the incredible dedication, knowledge and support of the non-professional staff, who deal with the switchboards, reception, secretarial duties, accounts, administration, refreshing supplies of stationery and the kitchen and a myriad other duties that are unremarkable in themselves but vital to the smooth running of any business. Many support staff will also have strong relationships with the clients supported by their partners and will therefore play a vital role in the delivery of a high-quality and personalised client service.

Sadly, firms often fail to appreciate (and express their gratitude to) support staff until they leave and the enormous range of tasks they undertook efficiently and without complaint remain undone. Support staff will often have been in the firm for a very long time and will know the firm's systems inside out and back to front. They will often have an insightful understanding of the strengths and weaknesses of the professional staff within the firm and ensure that the appropriate calls and tasks are allocated accordingly. They will often be the people who placate the clients and make up for lapses in the firm's service when professionals are busy or away from the office. As mentioned in the earlier chapters on marketing, the support staff are key "brand ambassadors" of the firm and an important element of the client service delivery process.

So please treat them with respect and say "Thank you" as often as possible. Recognise that they are intelligent people who know the firm and its people and its clients really well and can provide ideas and input that might save significant costs, help improve the service and generate new business. Avoid an "us and them" mentality. Ensure that they are invited to join team meetings so they are able to see the bigger picture. Encourage them to attend briefings and internal training sessions so that they understand the firm's strategy, develop their knowledge of the firm and thus can help clients more effectively. Invite them to client events so that they can seal their telephone and email relationships by putting a face to the name.

As with your professional staff, you should have clear job descriptions, a grasp of the competencies required to operate effectively and a sensible pay and promotion policy. There should be a training programme available to support staff and you should devote time and attention to their appraisals. It does not take much to keep your support staff trained, informed and motivated yet the benefits of having these important people onside and motivated are invaluable.

Checklist and actions

- Have you explained the firm's strategy to your support staff?

- Are all support staff aware of their contribution to the delivery of a quality service in line with the firm's values?

- Are support staff invited to briefings and team meetings?

- Have you explored how support staff can be more involved in developing new and existing business?

It's all in your mind — knowledge management

In a small firm, knowledge management is not a great issue — you can walk over to every member of the firm and ask a question, you can hear the telephone conversations they have with clients and you can have a detailed understanding of what they are doing and how. However, as your firm grows you will increasingly find it difficult to keep track of all the market knowledge, client knowledge and technical knowledge that resides within the heads of your people. As you grow, you will start to recruit specialists who have technical knowledge that no one else in the firm has. Do you really want all that knowledge to just walk out the door when your staff leave the premises?

The economic prosperity of a firm depends on the effective exploitation and retention of its organisational knowledge. Peter Drucker introduced the concept of the knowledge worker in the 1960s and he referred to knowledge replacing capital, natural resources and labour as a basic economic resource. Davenport and Prusak (1998) defined knowledge as a "fluid mix of framed experience, values, contextual information, and expert insight that provides a framework for evaluation and incorporating new experiences and information".

Distinctions must be made between data, information and knowledge. Knowledge is familiarity, awareness or understanding gained through experience or study. It results from making comparisons, identifying consequences and making connections of information. Some differentiate between explicit knowledge (formal models, processes, rules and procedures which can be communicated externally) and tacit knowledge (mental models, experiences, stories, rituals and skills residing in the individual and private mind).

Knowledge management is defined as "applying the individual and collective knowledge and abilities of the entire workforce to achieve specific organisational objectives" (Fong and Cao, 2004, RICS Foundation). They argued that the goal of knowledge management was not to manage all knowledge, but rather to manage knowledge that is important to an organisation.

Where you are providing a similar service to many different clients, or where you have lots of clients with a similar problem the opportunity for "reinventing of the wheel" starts to increase and the firm can become both inefficient and inconsistent. Furthermore, if you captured critical "intellectual capital" it could be packaged up to make methodologies or innovative new services. The surveying profession has a long history and is knowledge-intensive in nature. Table 8.2 explains how using knowledge improves the service provided.

Table 8.2: The service-knowledge spectrum

Service as service	Service as knowledge
Carried out for clients	Enhances client's knowledge or competence
Saves time and effort and creates convenience	Initiates improvement and development and creates change
Task-oriented	Process-oriented
Uses competence internally	Transfers knowledge
Aims to take over responsibility	Aims to help clients to help themselves
Packaged "product"	Adapted to client's unique needs
Costs	Investments
Repetitive	Creative/innovative
Runs operations	Runs projects

Source: Revised from Nicou et al 1994

Knowledge management is a relatively new management specialism. It has three basic elements: people, technology and processes. People create, share and use knowledge. Managing knowledge consists of deciding with whom to share, what is to be shared, how it is to be shared and ultimately sharing and using it. The success of knowledge management initiatives depends on people's motivation and their willingness to share their knowledge and use the knowledge of others. You need to explore cultural support for reciprocity (ensuring that if some provide their knowledge, others will do so also), repute (ensure the right to be seen as an expert remains) and altruism (personal satisfaction from sharing knowledge and helping others).

Processes to acquire, create, organise, share and transfer knowledge are characterised as follows (Nonaka and Takeuchi):

- Socialisation — the direct conveyance of tacit knowledge through shared experience.
- Externalisation — articulating tacit knowledge into explicit concepts via models, hypotheses and standard processes.
- Combination — systemising concepts into a knowledge system.
- Internalisation — embodying explicit knowledge into tacit operational knowledge (ie, training).

Technology stores and provides access to knowledge. There are some very sophisticated theories and software to tackle knowledge management in the largest professional practices, but simple (and almost cost-free) approaches are available to smaller firms. Email can be easily used to request knowledge from others on particular matters or to alert others to the existence of some new knowledge that they might find helpful. The intranet or other groupware can be used to set up pages for different teams and departments to deposit key documents and information. Ensuring that team meetings have a time slot allocated to talk about new developments or potential new services that can be packaged (see chapter 4 on service and product development). Furthermore, developing standard practices where, for example, every time someone attends a course they provide a short summary to everyone else in the team and place the core elements of what they have learned onto the intranet is an effective way to promote the sharing of knowledge. Some firms — and we see this particularly in the legal sector — appoint "professional support workers", whose sole job is to collect, organise and help others deploy the knowledge of the firm.

Common knowledge management practices include: knowledge audits to determine and locate the knowledge that is needed, knowledge maps to allow quick access to the knowledge, groups established to share knowledge and train new members of the team, documentation of best practice and feedback, content management to keep knowledge current and relevant and story telling to convey knowledge. The leading professional service firms invest a great deal in their knowledge management processes — and they can provide competitive advantage and additional value to clients.

For example, in residential agency, each agent has a huge wealth of knowledge about the area, the types of residences, each locality, the local providers of finance and insurance and legal advice, the nature of the buyers, the Council search and planning processes, the local schools, the traffic and so on. If all the agents could pool the key elements of their knowledge and make it accessible to all other agents in the firm, every agent would be more effective. The firm would also have a resource that it could draw on to improve the service package to all house buyers and thus differentiate the service from other local agents. Similarly, consider the issue of commercial valuations — there are standard rules for how these must be done, but if a firm could convince its most expert valuers to articulate and record their specialist knowledge it would be possible first to create a standard methodology for all commercial valuations undertaken by the firm and also it will be easier and faster to train new valuers joining the team.

Knowledge management is a fast-growing area and there are some excellent books available — although most are focused on legal firms the principles are the same.

Checklist and actions

- Is there someone in your firm who is tasked with learning about knowledge management and how it might help your firm grow or achieve competitive advantage?

- Have you identified those areas of knowledge that are critical to your firm's current operations?

- Does the culture of your firm promote the sharing of valuable knowledge?

- Does your intranet have areas where different practice groups can store, structure and access important knowledge?

> - Are there areas of practice where your firm has some special or unusual knowledge that could be collected and packaged to create a new (innovative) service that differentiates it from the competition?
>
> - Could you improve the efficiency and quality of certain services by encouraging the relevant professionals to pool and structure their knowledge?

Succession

While this topic is near the end of the chapter, it is probably one of the main concerns of partners in property practices and one of the greatest limitations on a firm's future growth. The reason for its late appearance in this chapter is because the first step in ensuring you effectively grow the right leaders for your firm in the future is to ensure that you have paid full and proper attention to all aspects of human resource management that will support the overall processes needed to recruit, retain and develop the right calibre of staff that will be the candidates for your next generation of partners. Developing leadership skills is also touched on in chapter 3.

You need to take a long-term view of the business as part of your business planning process to work out not only how many partners you will need in the future, but also the nature of those people and the skillsets that they have. And the profile for your firm's future leaders might be very different from the profile of those who currently run the firm. Also, you need to accept that the markets and the firm are changing rapidly and this change will only accelerate. Therefore, what makes an excellent partner today is unlikely to be the same type of person who will run the firm in the future. Motivation is also an issue — your present partners accept the present risk-reward profile of equity partnership but the current young generation of surveyors and agents may have very different expectations, particularly with regards to the work-life balance, which is much more important to today's youngsters than it was when we were young. The following work of Zemke highlights these generational differences:

- Veterans (born before/during the Second World War)
    - Attracted to workplaces with stability and who value experience.
    - Loyal to employers.

- Boomers (born 1940s–1950s)
  - Place high value on effective employee participation.
  - Do not object to working long hours.

- Generation X (born 1960s–1970s)
  - Enjoy ambiguity and are at ease with insecurity.
  - Require proper "work-life" balance.
  - Resistant to tight control systems and set procedures.

- Nexters (born post-1980)
  - Wholly intolerant of all unfair discrimination.
  - Serious minded and principled, prefer to work for ethical employers.

Another issue to consider is that planning for effective succession in terms of a retiring partner where a number of important client relationships must be handed requires several years' lead time. The problem is exasperated as the retiring partner will wish to keep these key client relationships to his or herself until the last possible minute. But to guard against client defection you need to ensure that younger professionals establish effective relationships with the next generation of the client organisation's leaders — so these are different relationships to those managed by the retiring partner. There is more information on this in chapter 3 on clientology.

Checklist and actions

- Do you have a firm-wide plan for succession for those partners who will retire in the next five years?

- Have you built a profile of the competencies and qualities needed by future partners?

- Have you identified how you need to develop those professionals with the potential to succeed?

- Does each of the older partners have a plan for how and when to hand over clients when they retire?

And don't forget ...

Human resource management and employment legislation are huge topics and we have only scratched the surface in this short chapter. There are many other issues of which you should be aware and it is best for you to seek expert advice in many of these areas. This chapter alerts you to some of the most critical issues.

Stress

Recent cases have shown that employers can be considered negligent (and liable to pay significant damages) if they allow staff members to suffer unreasonable levels of stress. Psychological research shows that people react very differently to stressful situations and that what is a perfectly happy situation to one person may be the cause of acute stress to another. To make matters worse, stress is a difficult concept to define and the Courts indicate that if a member of staff "perceives" that they are stressed, then that is sufficient for a case — regardless of the reality.

It should also be noted that management teams can be responsible — inadvertently — for causing stress and the frequent cause is a lack of clear and regular communication with staff (see internal communications above). Common management actions that increase stress include: announcing changes piecemeal, allowing staff to learn about changes through the rumour grapevine or the media, refusing to explain reasons for change, not providing a vision of where the firm is heading, increasing uncertainty by long delays between announcing changes and advising those directly affected and dishonesty. To reduce stress, management teams should try to see things from the viewpoint of those affects (empathy again!), involve and consult with those affected at the earliest opportunity, provide support to those affected, treat all concerns (both real and perceived) seriously and communicate at every possible occasion.

Contract of employment

All members of staff will be required to sign a contract of employment. Not only will this important legal document set out the terms of the employment — the pay, benefits etc — but also it should state the minimum standards of behaviour expected of employees in return. It might also tackle professional standards that must be adhered to and/or legislation relating particularly to the property industry (eg, estate agency and financial services laws) and the professions (eg, money laundering). It should deal with issues relating to grievances and the termination of employment — particularly issues such as confidentiality, copyright and client and staff "poaching" (restrictive covenants). Partners or directors may be required to sign a service agreement or a partnership deed. These vitally important documents must be checked on a regular basis by qualified lawyers.

Employment legislation

Employment legislation is a minefield of complex laws that has undergone radical change over the past few years as a result of European laws and key cases in the UK. It is too broad a subject to expect anyone other than a legal expert to remain up to date, so you are urged to seek out an expert to help you and your firm avoid the numerous pitfalls.

To illustrate, below is a short list of some of the most important pieces of employment legislation, although there are many more. Many property firms will have contacts within local law firms who are likely to provide regular briefings and seminars to review these and new developments for free and I would urge you to take them up on their offer. It will also provide you with an opportunity to identify suitable employment lawyers to work with your firm should the need arise.

* Equal Pay Act 1970;
* Health and Safety at Work Act 1974;
* Rehabilitation of Offenders Act 1974;
* Sex Discrimination Act 1975;
* Race Relations Act 1976;
* First Aid Regulations 1981;
* Trade Union and Labour Relations (Consultation) Act 1992;
* Employment Relations Act 1999;
* Disability Discrimination Act 1995;
* Asylum and Immigration Act 1996;
* Working Time Regulations 1998;
* Part-time Workers Regulations 2000;
* Fixed-Term Work Directive (Employment Act 2002);
* Equality (Sexual Orientation) Regulations 2003; and
* Employment Equality (Religion and Beliefs) Regulations 2003.

Other sources of help

The Chartered Institute of Personnel Development (CIPD) is the professional body for human resource specialists. The website contains a wealth of valuable information and there are a large number of open courses provided to tackle particular topics. It also publishes a series of easy-to-read guides on common human resource issues, which are worth reading — I have listed some of the most useful books in the reference section.

Case study: Brown & Co — people

Jim Major became the managing partner of Brown & Co in 2003. While continuing to spend two-thirds of his time in active client management (he is a rural agent based in King's Lynn), he has been busy leading the development of the practice.

Changing management structure

When I started in the management role, deciding what to do was relatively easy. During my first year, I had a list of some 30 actions. For the second year this was down to just six, although most of these were the tougher projects to tackle. Brown & Co has always wanted to grow and develop and there was a general feeling among the partners that there was "unfinished business" in terms of what we could achieve. We were essentially a firm of individual practices — each partner (many based in geographically different offices) had an individual fiefdom and we liked the freedom and autonomy it gave us. We weren't sure what we needed to do next but knew there was work to do. As "amateur managers" we recognised a need for professional help — to engage us in our thinking and to help us look at our situation and its challenges and opportunities in a proactive way, with the benefit of experience of other partnerships at a similar stage in their life.

It was reassuring to know that Brown & Co's development was typical of that of a professional practice and that we "fit the model". I am most proud of the way in which we have structured the practice now. There is an overall plan detailing the firm's overall objectives, then divisional plans, which draw together the rural, commercial and residential practices across all 10 offices and then individual office development plans. It feels more efficient now and we can see movement towards the achievement of our goals. We are co-ordinated and focused.

Most medium-sized, growing practices probably need some sort of crisis to drive the type of management change we have seen here at Brown & Co. However, we have had 10 years of good growth and two to three years of even better years behind us. As a firm we are not complacent and we like to encourage initiative among our partners and staff — that's how we extended the practice into Poland and St Lucia. We are ambitious but not greedy and understand that we need to invest to build the practice further in the future.

Developing future talent

While focused on developing a strategy and management structure to take the practice into the future, we were also keen to look at ways to further engage the young surveyors in management. We have a Development Forum where associates from around the firm come together on a quarterly basis to tackle real management issues on behalf of the management committee — they have produced excellent ideas and programmes in the areas of market research, competitor analysis, staff bonuses, developing relationships with banks and other intermediaries and training.

In addition to providing them with a chance to participate in management, it has also helped us accelerate their development and admit some into the partnership — so it's supporting our succession plans. It is interesting also for the partners to understand what younger members of the firm think about its future direction.

Hiring professional managers

We were already convinced of the benefits of hiring professional managers — our HR manager was previously a recruitment consultant and has helped us streamline our recruitment and training processes and save considerable costs. Our IT support we have outsourced — and receive a responsive and comprehensive service as a result, at a lower cost than employing in-house.

We have also recently appointed a senior marketing and business development professional. We knew that we needed to extend our marketing beyond our successful newsletters, county shows and media relations programmes. Having spent time engaging the entire partnership in a strategic review of our markets — which included research with our clients — it became clear we needed a senior manager to spearhead development in this area and to bring more focus and sophistication to our efforts. We are investing more in this area than we originally anticipated but are confident it will produce the results we want.

Leadership

Partnerships are, by definition, about all partners having equal responsibility and input into the firm's development. In the role of managing partner, you inevitably have to adopt a leadership role with an emphasis on delivery and decisiveness — with more time to invest in analysing the problems and understanding the background you develop a better view of the problems and possible solutions. Yet I have learned that humility is probably the greatest asset for any leader in this environment and a great help is experience in sports — you may be captain but you are still very much a part of the team.

Brown & Co (*www.brown-co.com*) is a Limited Liability Partnership with 19 partners assisted by over 145 professionals and support staff working from 10 offices throughout East Anglia and the East Midlands, and on an international scale with offices in Poland and St Lucia.

Change management (flexible strategies)

Roadmap

- Why is change so difficult?
- Cultural context — hawks and doves, lone wolves and worker bees
- Understanding how people change
- Are they with you?
- What's in it for me?
- What type of change?
- Planning for effective change
- Scream if you want to go faster
- Leading change
- Managing organisational change
- Communication
- Do as I say, not as I do — partner role models
- Delegate and be done with it
- Crisis — what crisis?
- Project management in a nutshell
- Time management basics
- Merger mania
- Using outside consultants
- Change fatigue — not another change initiative!
- Don't always look up to the peak — celebrate your successes

"It must be remembered that there is nothing more difficult to plan, more uncertain of success, or more dangerous to manage than the creation of a new order of things."
Machiavelli

"Do one thing each day that scares you."
Eleanor Roosevelt

We are nearing the end of our journey into management. I hope you are still with me because this final chapter is probably the most important. You may have a plan, you may be a great leader, you may know what to do to sort out your strategic and tactical marketing, improve your selling and enhance client service and retention. Now all you have to do is change the way that the partners think. Change the culture, change the practice, and thus your problems begin.

After navigating the perils of goal setting in a partnership and preparing a hefty document, many partnerships feel that they can tick the "planning" box as the ink dries on the various plans and promises. Real life returns, client demands build up and no one ever tackles the changes that you have all so carefully researched, sweated over, fought over and finally agreed.

But achieving change — in attitudes first before you can really change the resulting behaviour — is almost a science in itself. There are many tools and techniques, but you must be careful to choose those that are best suited to your own personal style as well as to the nature of the people — and the culture of the organisation — that you are trying to change. This relates back to the earlier material on leadership — so now you must lead the change.

Why is change so difficult?

Familiarity is comfortable. We all like to stay within the safety of our carefully crafted comfort zones. A few of us like the thrill of risk taking — but how about the prospect of shouldering the blame when your colleague might say: "See, I told you so".

Achieving change — both within the hearts and minds of your fellow partners and your staff — and thus at the very heart of your culture and ethos is perhaps the biggest challenge of all. It is dangerous — there is always the risk that you might accidentally or inadvertently lose the *"je ne sais quoi"* that got you and your firm to its present successful position.

It is tough — professional people are naturally risk averse and fearful of failure. You are now leading the firm — and the team. And they look to you for your strength, wisdom and guidance. How would they feel if you let them down? How would you feel if you make a wrong decision that put their jobs, mortgages and careers at risk? Changing things is scary. That's why you have read this book and tried to apply all the tools and techniques in a way to minimise the risk of making a bad decision.

Culture context — hawks and doves, lone wolfs and worker bees

You need to be aware of the impact of your firm's culture on any change. Some property partners consist entirely of "lone" individuals or wolves who are accustomed to working alone, in their own particular style without recourse to their fellow partners. Other partners are more comfortable as part of a "corporate" collective — like a hive of happy worker bees. They understand the role of their colleagues and are happy to share the load and have a division of labour during large projects. If you have mostly lone wolves, then change programmes will take much longer than if you have a colony of worker bees.

Then there will be the extent of their tolerance towards ambitious change or hard rules. The hawks will be aggressively seeking strong growth and will be intolerant of any one who is not seen to be pulling their weight. Whereas the doves are happiest when harmony is maintained and no one rocks the boat. The hawks will be driving change — noisily — and the doves will be sitting on the sidelines trying not to become involved. You need a blend of hawks and doves but the predominant theme in your firm will dictate the pace of change.

Understanding how people change

What is required if we want individuals to change? Derek Pugh, an expert on change, observed that the reaction of many people to poor performance or criticism is to do "more of the same" only harder. They stick to the familiar — their comfort zone. When there are threats or unknown things on the horizon they respond by psychologically "going rigid" — think about a rabbit stuck in the headlights of an oncoming car.

He suggested that in order to help people change they first need to be "unfrozen". This means that you must reduce the immediate threat and create some psychological safety for them. Reassure them and let them know that change is always scary and that they are among friends who understand how challenging it all is and that they want them to succeed. Once this feeling of safety is established, you can create the change or movement that you desire before helping them "refreeze" in the newly established patterns of behaviour. Pugh also observed that: "Individual behaviour is powerfully shaped by the organisational roles that people play", so think back to the material on leadership, management structures and the material in chapter 8 about job descriptions and being explicit about your expectations of them. He also noted that: "The least likely to understand and accept the changes are the unsuccessful". That's one that I sometimes throw into partnership conversations when I meet particular resistance!

In the psychology world, there is much material on how people change. I particularly like the thoughts about transition explained by William Bridges. He recognised that people have to "mourn" the end of "the way things were" and probably need some time out when these negative emotions ran their course before they were ready to move forward to the new shape of things to come.

- **Endings**
  - Disengagement
  - Disidentification
  - Disenchantment
  - Disorientation

- **The neutral zone**
  - Time out
  - Disconnected
  - Frightened

- **The new beginning**

Other psychologists have observed this change cycle and seen other aspects of the process — some have even identified where people are most likely to have negative energy (moans, complaints, resistance) and most positive energy.

So what this means is that change is never simple. And that, to some extent, resistance and anger and even depression are natural and inevitable. Knowing that each person will need to go through the cycle — and at different rates and with a different degree of severity — does not make things any different but should provide you with reassurance and comfort that other firms will experience the same challenges when they embark upon similar change programmes.

Figure 9.1: The change cycle

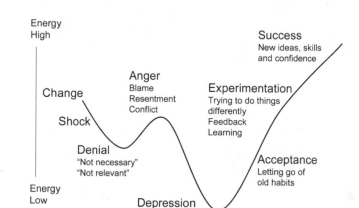

My resistance is low

It's a great song isn't it? Jane Russell and Hoagy Carmichael at their best. I love the lyrics as it must be the wish of any consultant and change manager — going into a firm and hearing all the partners and staff singing "My resistance is low". Here I want to provide you with another short checklist of things you can do to overcome resistance. These are from Cotter and Schlesinger.

- Education and communication.
- Participation and involvement.
- Facilitation and support.
- Negotiation and agreement.
- Manipulation and co-operation.
- Explicit and implicit coercion.

The sections below deal with these in more detail.

Checklist and actions

- How ready is your firm for a major change programme?

- Are your partners (and staff) committed to the proposed change programme?

- What is the mix of hawks and doves, lone wolves and worker bees — how will this affect the planned change(s)?

- Are you prepared to modify the pace of proposed change to enable your people the necessary time to adapt?

- How much resistance is there to the planned changes — have you explored all the possible issues?

- How will you overcome resistance from different partners and staff?

Are they with you?

An expert in the area of individual change noticed that "The relative strength of an individual's identification with and involvement in an organisation" is a critical factor in the extent to which they roll up their sleeves and get down to the business at hand and changing themselves and thus the firm. Three components of this were identified by Griffin and Bateman:

- The desire to maintain membership of the organisation.
- Their belief in and acceptance of values and goals of the organisation.
- Their willingness to exert effort on behalf of the organisation.

These topics are covered further in chapter 8 on people. You might want to consider those around you and think about which of the following categories they belong to:

- Defiance — Not on board. You still have a major persuasion job to do with these ones. But be careful because if they are very senior and/or powerful or influential in the practice their defiance may spread to others!

- Compliance — Grudgingly coming round. These people are doing what you asked but not always with the enthusiasm or passion that you had hoped. Encourage them for they are trying but ensure that you give some extra attention to win their hearts as well as their minds.

- Alliance — With you all the way. These are your champions. The eager beavers who know what and why you are trying to change things and are not only doing what you asked of them but are actively trying to encourage others to get on board too.

An alternative way of thinking about what "sides" people are on is to consider whether they are pretenders, leaders, dissenters or followers, depending on their agreement with the change and also their effort to implement.

It may help here to think about the "WhereRU mindset" (John Frost of *Values Based Leadership*). He advised that you should identify people's mindset when they face enforced changes in their work life so that you can guide them from a position of disempowerment to one of empowerment. The tool consists of a visual reference graphic and is supported by questioning techniques.

Figure 9.2: WhereRU mindset model

3. Realistic Accepting + Constructive?	4. Responsive Active + Collaborating?	**Empowered**
1. Resentful Angry + Complaining?	2. Resigned Ambivalent + Criticising?	**Disempowered**

Source: John Frost, Values Based Leadership

Organisational commitment has only very loose links with overall job performance. However, highly committed people are more likely than less committed people to help others in the organisation to effect the proposed changes. These are the good guys (and gals) that you need around you when embarking on a change programme.

> "Champions are individuals who take on an idea (theirs or that of an idea generator) for a new product or service and do all they can within their power to ensure the success of the innovation by actively promoting the idea, communicating and inspiring others with their vision of the potential of the innovation. Champions help the organisation realise the potential of an innovation." (Afuah, 1998)

What's in it for me?

In chapter 8 on people we touched on the issue of motivation, reward and recognition. There is little point in asking (or telling) people to change if the systems in place are still geared up to measuring and rewarding the old ways of doing things.

Having spent long hours and sleepless nights analysing the situation, talking earnestly to your partners about the various options and choosing (nervously) the right approach to take you think that a few words to the troops — over a glass or two of wine — is enough to galvanise them into action.

Communication is really important. So is involvement. You have to help your people to see what you saw, they need to understand the issues and options and they need to see how the change programme — especially if it is dramatic or will take a lot of time and effort — will benefit them as individuals. You need to show people that the change will be good for the firm and directly how it will help them — more money for bonuses, more interesting work, better clients, opportunities to work abroad, space within the equity partnership, greater job security or a more comfortable work environment.

What type of change?

Knowing what type of change you are advocating will provide insight into what is likely to happen — and the probable reactions. Some change is evolutionary and incremental — gentle and safe and background. Other change is more radical and revolutionary — and this is the scariest stuff.

The type of change will also suggest the type of resistance that you are likely to face — and also the appropriate ways of dealing with that change (see figures 9.3 and 9.4).

Figure 9.3: Leading different types of change — incremental

Urgency \ Resistance	Low	High
High	Extensive participation	Persuasive
Low	Focused participation	Persuasive or coercive

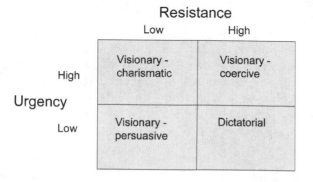

Source: DE Hussey

Figure 9.4: Leading different types of change — fundamental

Urgency \ Resistance	Low	High
High	Visionary - charismatic	Visionary - coercive
Low	Visionary - persuasive	Dictatorial

Source: DE Hussey

Planning for effective change

As with most things in life, a little advance planning can make such a big difference — saving time, tears and frustration.

There are a number of frameworks here that I have found useful working with property partnerships. For example:

Phases of planned change (Bullock and Batten)

- **Exploration phase**
  - Whether an organisation wishes to make specific changes and commit resources to planning the changes.
  - This relates to the point above about giving people a proper chance to think through the proposed changes, consider all the pros and cons, identify the specific tasks and consider the impact on themselves and others before they embark on the change.

- **Planning phase**
  - Understand the problem. Collect information, establish change goals and get decision makers on board.
  - This should ring bells as, in the majority of planning frameworks in this book, we have considered the need for people to collect the relevant information, immerse themselves in analyses, make comparisons among various options and set a course that seems appropriate.

- **Action phase**
  - Implement the changes.
  - Everyone needs to know exactly what they are supposed to be doing and why. People are not telepathic — they cannot read your and the partners' thoughts. You need to tell them in simple terms what you want. Then ensure that these requirements are backed up in job descriptions, embedded in the reward and recognition systems and set out in a clear plan — that is communicated.

- **Integration phase**
  - Consolidate and establish the changes so that they become part of normal, everyday life.

- Once the project has got under way and things move in the right direction, make sure you tell people about the successes, thank them for their efforts, remain alert to their concerns about current and future difficulties, be prepared to revise some of the parameters or requirements of the change and help those changes bed down and become part of "the way we do things around here".

Set out the objectives, tasks and methods for each stage of the change programme and consider what the appropriate outputs you might expect to see on the successful completion of each stage. See the material below on project management.

Analysing the forces working for and against change

Before you embark on a major change programme, it is worth undertaking a systematic study of the general forces for and against the proposed change. This is because while there may be sufficient enthusiasm to embark upon a change programme, there may be many barriers and obstacles getting in the way. Having a greater knowledge of what might rise up and bite you in the bum as you launch your exciting new initiative might just give you a bit of comfort. After all, forewarned is forearmed.

You can, also, simply write down a list of the pros and cons of a particular change programme. And ask your colleagues to talk through with you what is likely to happen — to add in their ideas and to try to consider ways of preventing the bad things and encouraging the good bits.

Sometimes, it helps to draw a diagram and see the situation visually. A force field analysis can also be instructive. Here's one I prepared earlier (figure 9.5).

Figure 9.5: Force field analysis

Source: Lewin

Checklist and actions

- What is the mindset of the various people involved in the change programme?

- Who are your champions? Who is most likely to remain defiant and who will comply?

- Have you taken the time to explain "What's in it for me?" to the various groups of partners and staff according to their personal aims and motivations?

- Have you considered what type and pace of change you are proposing and the best way to lead?

- Have you done sufficient planning — and allocated time and energy for the different phases?

- Have you completed an analysis of the forces for and against change — and the potential barriers and obstacles?

Scream if you want to go faster

Just as we saw previously that individuals each progress through their own change cycle with all its attendant highs and lows, so too does an organisation, as figure 9.6 shows.

Figure 9.6: Transition steps

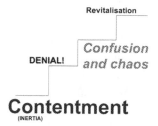

Source: Ashridge College

Hare or tortoise? What is your firm like? The history and culture of your firm will dictate just how fast you can safely navigate the planned change. Take a look at figure 9.7 and consider the route and pace that are realistic for your firm.

Figure 9.7: Change management

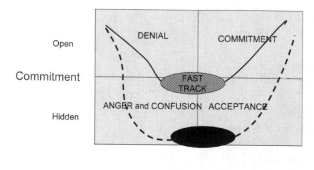

383

Leading change

Being the architect of the plan is not sufficient. You now have to be the leader who takes each and every member of your firm (sometimes kicking and screaming) through the change programme. Refresh and remind yourself of some of the tools and ideas we covered in chapter 3 on leadership, you will need all the help you can get now!

In that earlier section on leadership, we explored various models of leadership and the likely behaviours of successful leaders. In the book *Leadership for absolute beginners*, there is much advice for leaders on how to bring about change, which can be summarised as follows:

1. Change the organisational culture.
2. Raise people's awareness about rewards.
3. Help people look beyond self-interest.
4. Help people search for self-fulfilment.
5. Invest others with a sense of urgency.
6. Commit to greatness.

Might it be easier if you had a cute little acronym to remind you of what you need to do? Here is one that I like by Hussey to make it EASIER:

- E — Envisioning. Creating the vision of where you want the firm to go.
- A — Activating. Generating enthusiasm from everyone to take the journey.
- S — Supporting. Helping everyone to know what needs to be done and how.
- I — Implementing. Giving guidance on what is to be done, when and how — in bite-size pieces.
- E — Ensuring. Checking that the plans are achievable and realistic and that if real problems are encountered that you will deal with them.
- R — Recognising. Noticing (and thanking) people who are really working hard and making the vision a reality.

Interestingly, the first three of these steps need your leadership qualities of charisma and behaviour. The latter four rely more on management and administrative skills. So now that you know your personal leadership strengths and weaknesses, perhaps get some of your partners with complementary skillsets to lend a hand too.

Managing organisational change

I know that you do not want any more on management stuff (that is why you bought this book) but there is a great book called *How to manage organisational change* by D E Hussey (published by The Sunday Times). It is short but full of good ideas and sound advice on managing changes within an organisation.

There is research (Larry Alexander) as to why organisational change often fails and the main reasons being:

- Implementation took longer than planned.
- Major problems surfaced after the project began.
- There was inadequate co-ordination.
- Competing activities and crises diverted attention.
- Managers and people lacked the capabilities to implement the changes.
- Inadequate training and staff instruction.
- Uncontrollable external factors.
- Poor leadership from departmental/team leaders.
- Key implementation tasks not defined.
- Information systems inadequate for monitoring implementation.

Hussey's "integrated organisation" is a good model that integrates all the elements needed for an organisation to successfully navigate a major change programme and it provides a helpful checklist of things to address before embarking on your change programme.

Figure 9.8: The integrated organisation

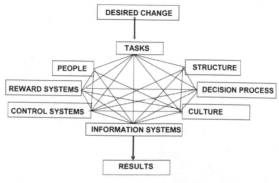

Source: DE Hussey

Communication

A great many books have been written about effective communication for business leaders. You should also look at the communications material in chapter 8. Some of it is great, some of it is terrible. One piece of advice that I really like is from Jim Collins in his book *Good to great*. He talks about the need to create a climate where the truth is heard, and he offers the following advice:

1. Lead with questions, not answers.
2. Engage in dialogue and debate, not coercion.
3. Conduct autopsies, without blame.
4. Build "red flag" mechanisms.

They are all fairly obvious apart from the last one. Sometimes, when you are the leader people are afraid to give you bad news. They might tell you that all is well, even when it is not. Things such as your change management programme might be going seriously awry and everyone knows it except you. So his advice is about identifying those things that you should watch carefully that will tell you — when no one else will — when things are going wrong. The other great advice he gives is a reference to The Stockdale Paradox. He explains that is necessary to retain faith that you will prevail in the end, regardless of the difficulties and at the same time confront the most brutal facts of your current reality, whatever that might be.

Unfortunately, the amount of attitude change is directly related to the degree of attractiveness of the change agent (Tannenbaum). The credibility of a communicator rests partly on his or her expertness and trustworthiness as perceived by the person on the receiving end of the communication (Hovland and Weiss). This means that if people find you unpleasant, then there is little likelihood that you will be an effective communicator. Should this unlikely scenario be the case, fear not, for help is at hand. The "sleeper effect" is where the person remembers the message but has forgotten the source, which means that you can use your colleagues to plant ideas in people's minds.

Checklist and actions

- Have you considered how the firm will manage as you progress through the change cycle and some people become confused, impatient or upset?

- Have you decided on a fast or slow pace for the change programme?

- Do you know how best to lead everyone?

- Have you thought about the stress levels for everyone in the firm and what you can do to reduce stress?

- Have you developed a comprehensive and ongoing communications programme to support all phases of the change programme?

Do as I say, not as I do — partner role models

I have often found that the partners develop a great plan and then look to the less senior surveyors and agents (or even the poor downtrodden and overworked support staff and secretaries) to make the change happen. I know a number of large and respectable firms that have crafted great plans and strategies and then invested enormous amounts of money on internal communications and training programmes to help all their people understand the planned change, what part they are expected to play in achieving the change and the new behaviours that are required but have wasted all the money. Why? Because people in the firm take their lead from the most powerful and influential roles models that exist — you and your partners.

If you want to achieve change then you must start at the top. A concentrated burst of energy directed at getting your partners to start exhibiting different behaviours will pay dividends in terms of how effective this is at creating a real sea change among all those who work for those partners.

There have been many occasions when I have been very impressed by the associates and assistants in firms. They are often well versed in management (it is now incorporated into elements of their professional training), more commercially aware of what is going on in the big wide world around them, confident in their use of technology, how the new online world can be used to the firm and clients' advantage and full of enthusiasm for taking on new challenges and tasks to help their traditional, and perhaps underperforming, firm move into the 21st century. But often these young people have been blocked by the partners who either do not understand or are wary of new things. So the youngsters put away their enthusiasm and emulate their partner role models.

Delegate and be done with it

It may occur to you as you fall down in an exhausted heap that you are simply too tired to do it all on your own that you should delegate. Unfortunately though, the property professions — as with most other professions — seem to be particularly bad at delegating.

Anyway, delegate away. But follow a few simple rules (these are from a book called *The art of delegation* by Ros Jay and Richard Templar) if you are to avoid disaster:

1. Review the task and set the objective.
2. Decide to whom you will delegate (and check that they have the appropriate authority, skills and time).
3. Set parameters:
    a. the objective (SMART right?);
    b. the deadline;
    c. the quality standards;
    d. the budget;
    e. the limits to their authority (exactly what they can and cannot do); and
    f. the details of any resources available.
4. Check that they understand.

I would also add another point here. Check back with them regularly to ensure that everything is all right or, better still, ask them to check back regularly with you.

Crisis — what crisis?

Often change does not happen because people are just too comfortable and they are complacent. There is no real need to change. Today, they have good work with great clients and generate healthy profits — so why bother with change? When things are going well, there is little incentive to focus seriously on change.

The recent awful events in the property market have shown one thing. When demand falls and things become tough, firms are forced to grasp nettles and deal with situations that have been left to fester for many years. The gravity of the situation means that tough and urgent action is required — and that it is taken. It is sad that it requires a firm to be in such a perilous position before it tackles many of the changes that it should have pursued in better times with perhaps a more considered and sensitive manner — rather than the "slash and burn" that we have witnessed in many firms recently.

The fact is that many firms for a long time were suffering "death from a thousand cuts" — there was no real crisis but their client base, or high-calibre staff base or their profits — where being slowly eaten away, leaving the firm less and less healthy. But none of these tiny cuts was enough to generate the "crisis" or "emergency" or "attention" discussions in the boardroom.

When the present difficulties lift, and they will, the surviving firms will start, once again, on their glorious growth with their confidence renewed and their memories of the bad times carefully repressed. But perhaps we should remember the power of the crisis to convince people to make hard decisions and embrace change rather than flirt with it.

If necessary, consider how you might simulate crises in the future. Run occasional exercises at board or partner meetings when you run scenarios where some awful situation must be considered — the loss of a major client, an aggressive move by a competitor, a cheap new online solution to one of your core services ...

Project management in a nutshell

The property and construction industries have their fair share of qualified and experienced project managers. In fact, for many firms — particularly quantity surveyors — this is a vital service that generates significant fees and profits. Why then is it that property partnerships rarely use this wealth of skills to help them manage the implementation of major change projects internally?

I do not have the time nor the space to provide chapter and verse on project management. But you need to know that is more than just a wish and a prayer. Really good project management (the sort that ensures the project happens on time, on budget and on spec) is almost a separate management discipline — with its own language, tools and methods. I use a relatively simple framework for smaller projects, and this is shown in figure 9.9.

Figure 9.9: Overview — simple project management

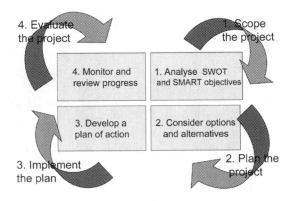

The principles of this planning cycle should be very familiar to you by now. You start by carefully analysing the situation and looking at the strengths, weaknesses, opportunities and threats. Then set some objectives (hopefully, this will also involve you looking at the cost-benefit analysis). Then you need to scope the project — decide where it starts and ends.

A word of warning here — many projects fail because they keep growing and growing because no one sets out clearly at the beginning exactly what would be involved. The problem with ever growing projects is that they have completion dates that never arrive. Now that you know what you need to do, you can consider the various options and alternatives available to you. Inevitably, your choice of strategy will depend on how serious the situation is and the time and resources available. But thinking through all the options and alternatives at this stage will prevent some bright spark sabotaging your project when it is half-way through because you had not given yourself enough time to come up with that very same idea.

At this stage, you can start planning out what needs to be done —
and remember to allocate the various tasks to specific people and give
them real deadlines. The messages about communication above should
be observed (tell people what is happening and why — and how it
affects them — and what you expect of them) and keep communicating
as the project rolls out. You also need to ensure that you allocate
regular time slots to review progress and keep things moving along.
Allow a healthy chunk of time during the life of the project for chasing
up slothful partners and the inevitable crisis management.

For more complex projects you need to do some more serious
project management. The five stage process is as follows:

1.   Initiate
     −   Identify stakeholders.
     −   Agree that the project is worth doing.
     −   Assess risks.
     −   Determine goals.
     −   Determine results.
     −   Define expectations for each stakeholder group.
     −   Define project scope.
     −   Develop a SOW (Statement of Work).

2.   Plan
     −   Refine project scope.
     −   List tasks and activities.
     −   Sequence the activities.
     −   Develop a workable schedule.
     −   Get the plan approved.

3.   Execute
     −   Lead the team.
     −   Meet the team members.
     −   Communicate with the stakeholders.
     −   Firefight and resolve conflicts.
     −   Secure the necessary resources.

4.   Control
     −   Monitor progress and deviations from the plan.
     −   Take corrective action.
     −   Receive and evaluate project changes.
     −   Reschedule the project if necessary.

- Adapt resource levels.
- Change the project scope.
- Return to the planning stage to make adjustments.
- Document and gain approval for changes to plans.

5. Close
   - Shut processes and disband the team.
   - Lean from the project experience.
   - Review the project's process and outcomes.
   - Write a final project report.

Believe me, you really need some help when it gets to preparing statements of task, network diagrams and Gantt charts (a chart that shows what is happening over a period of time). There's some great software around to support you in this too. But my advice is — if the project is really that big, hire in some expert help. It will be money well spent.

Time management basics

A critical part of your role as leader and project manager will be how you manage your own time and that of others around you. This is especially important if you expect people to get a significant number of hours down on their time sheets and still find time to do marketing, business development, recruitment, training, administration, team management, as well as tackling the major projects that you are now thrusting towards them.

As part of the International Professional Management Week, I produced a management diary. It is designed to help senior and managing partners monitor their aims and time use during one week so that they can assess where and how they might use their time more efficiently. It was also designed to show the difference between being busy and being productive. My friends in the consultancy world advise that you should always be looking at the outputs (results) rather than the inputs (time and activities).

I could write a whole book on time management. But instead I want to offer you some of the tools that I have found to be most effective with property partners. Are you ready?

Set goals

This is obvious. But you need to have clear goals if you want to be able to prioritise your time. Obviously, things that help you reach your goals take high priority and those that do not can be put aside (or delegated to someone else). The book by Ken Blanchard, *The one minute manager*, offers this approach to setting goals:

1. Agree on your goals.
2. Think about/see what good behaviour looks like.
3. Write out each of your goals on a single sheet of paper using less than 250 words.
4. Read and reread your goal, which requires only a minute or so each time you do it.
5. Take a minute every once in a while out of your day to look at your performance.
6. See whether or not your behaviour matches your goal.

I have seen a number of managing partners have the three or four sheets of paper on their desk that they regularly refer to in order to keep them focused on the big picture.

Allocate thinking time

Build time into your schedule to think or to deal with crises. Develop a pattern to your working day so that you are not constantly interrupted with emails, phone calls, or colleagues coming into your office for a chat. These are all things that need to be accommodated in your day — but make sure that you are in control about when.

Analyse what you need to do

I love the action priority matrix — it is one of my favourites (figure 9.10).

Recognise the difference between urgent and important

I recommend reading Stephen Covey's *Seven habits of highly effective people*. I particularly like his way of differentiating between urgent and important. Many partners get too diverted by the urgent stuff, so that the important stuff does not happen.

Figure 9.10: Action priority matrix

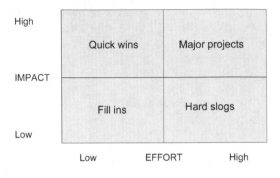

Figure 9.11: Urgent versus important

Source: Stephen Covey — Seven habits of highly effective people

Plan ahead

Consider what you want to achieve during the year and put milestones into your calendar. Then work out what you need to achieve each quarter, each month, each week and each day as a result. Start with the end in mind and work back. Otherwise, the danger is that your days will be filled with "stuff that happens" and you will not have time to do the "stuff that is needed".

Another neat tool here is to think about "rocks, stones and pebbles". Think of each day as an empty glass. Start by placing the biggest tasks (rocks) into the glass. Then fit in some of the medium-sized tasks (stones). Now you can fill the spaces around this with the pebbles and "sand" of daily life.

Procrastination?

Why do we procrastinate? It could be that other things are more comfortable, interesting or fun. Or maybe you do not know where to begin. It could be that you are fearful of failure or that you are a perfectionist and do not believe that you have the time to do justice to the project.

The important thing is to recognise when you are procrastinating, and then to try and work out why. Is it because the task ahead is unpleasant and you are avoiding it? Is the task too big and overwhelming? Does the reward seem insufficient for the amount of effort you need to make?

Then you just need to get over it! Some of the things that you might try include:

- Create your own rewards.
- Ask someone else to check on you.
- Think about unpleasant results of not doing it.
- Calculate the cost of the time to do it — and cost of not doing it.
- Break the project into smaller "bite-sized" pieces.
- Start with some small, quick tasks.

Other time savers

People in the property industry are incredibly polite. This means that often you will have to wait while someone natters on and on about a point when they really should move on. The more assertive of you will be able to tackle this. But there are some great ideas from Stuart Levine of Dale Carnegie & Associates:

- Watch non-verbal cues.
- Use "I got it!" and "You're killing me. Move on!".
- Close the loop — don't leave others hanging.
- Take time out from the Crackberry.

- Tell them if the baby is ugly — be specific.
- Organise yourself — On it. Pending. Done.
- Use PAC (Point Accepted. Continue).
- Good enough is good enough.

Checklist and actions

- Do your partners understand the crucial impact they have as role models?

- Are you and your partners prepared to lead from the front by adopting the proposed changes wholeheartedly?

- Do you know what and how to delegate?

- Are partners and staff sufficiently motivated to tackle the changes — or do you need to create a crisis?

- Is there sufficient project management expertise within your organisation or do you need to acquire some through training or external consultants?

- Have you devoted sufficient time and resources to building a good project plan?

- Do you have the necessary time to ensure that the project plan is monitored and managed adequately?

- Are you, your partners and your staff familiar with the most common time management techniques? Do you encourage their use?

- Are you procrastinating? Why? Get over it!

Merger mania

One of the most common results from business planning and strategic reviews and the like is the decision to go out and find either a marriage partner (ie, someone with plenty of money to buy you out or someone with equal fortunes so that you share risk and costs without diluting your profits), or you go on the hunt for suitable acquisitions — where you can bolt on specialist staff, make a splash in a new market or pick up a juicy client list. When you think about mergers and takeovers, you always focus on the upside. Time is rarely spent looking at the considerable downsides.

Most partners worth their salt with give the financial papers of the target company a thorough going over. Sadly, if this looks all right someone usually gives the green light to proceed. Wrong! There is a big "check out their culture" exercise to do, which means spending considerable time with the target firm's people working out whether you will be able to live happily ever after once the honeymoon is over. In addition, you need to undertake a lot of due diligence to make sure the target is as attractive as it first appears. Spend time looking hard at the client list — and the lost clients for the past few years — look at the number of new clients and transactions. Go through the CVs (and appraisals) of the key professional and support staff. Check that the leases and other contractual obligations are in order. Ask your IT people to give their systems the once over.

While it might seem like you are going to make big savings on the scale economies, make sure that you really will make a net saving on premises, support staff, technology and marketing once you take into account all the "transaction" costs. Also, I have seen several potential property partnership mergers wane when it comes to looking at the issue of partner pensions and annuities.

Even if you proceed, remember that there is a huge and ongoing amount of work to do to keep all the staff you want from both the new and old organisation happy and in sync. You will also need a programme to ensure that the clients of the acquired firm are properly informed about the changes and not alienated by your firm's different procedures and systems.

Using outside consultants

It is a fact that: "A prophet is not without honour except in his own land and home". This means that sometimes you need to pay an external consultant to come in and tell your partners and people what they do no necessarily want to hear from you.

When designing or implementing major change programmes, it is a common practice to hire help from external consultants. You will recall from numerous earlier mentions in this book that you need to have a clear brief for any consultants and that you need to trust them, take time to help them understand your organisation and possibly allocate them a couple of members of staff to help them find their way around your firm, obtain the information that they need and communicate effectively with your partners and staff.

Remember that a consultant can only do what you allow them to do. If a job is too tricky or too dangerous for you to manage, then the chances are the consultant will have trouble too. Remember the difference between delegation and abdication. A critical success factor for any consultant you choose to use will be how well they are received by your partners (and staff) and how well they understand and "fit" within your culture.

If the firm does not have confidence in the consultant, then they don't stand a chance of succeeding with even the smallest project. Invest some time during the selection process in checking that not only does your consultant have the relevant experience of the property sector and your location and the necessary specialist expertise to undertake the task at hand, but also the personality and style that allows them to win the hearts and minds of your fellow partners and staff.

Change fatigue — not another change initiative!

While change is good, sometimes firms become carried away and serve up change after change. The poor partners and staff have just been through one series of briefings and training sessions and started to assimilate some major new behaviour and you hit them with another change programme. And then another, and another.

As such, you need to think through all the potential change programmes that you are likely to need for the next two to three years. Make sure you select just the most important ones. Talk to your partners and staff about priorities. Assess the level of change that is palatable to everyone. Check that your support teams have the energy and wherewithal to keep up with you.

Schedule in times for consolidation. Where people can return to the comfort of normal day-to-day stuff without having to grapple with another new initiative. Reassure them that more change is on its way — but not today!

Don't always look up to the peak — celebrate your successes

Allied to change fatigue is the problem of always looking up the mountain — surveying the large number of new things that need to be completed and that are on the horizon. Take a look at your plans and the programmes for next year — impressive or scary?

For your own sanity and for the motivation of your whole team it is important, now and again, to stop and take a look behind you. Consider all the things that you have done and achieved. Spend a little while basking in the glow of a number of great jobs done well. And extend this small, self-congratulatory exercise to all those in the firm. It is immensely rewarding to be able to say — we set out to do a, b and c and we accomplished them all with some extras as well.

Take time out now and again to celebrate your successes. And look — you have just finished reading the entire book.

Checklist and actions

- Are you seeking a merger? If so, do you have a defined list of requirements that meets your overall strategy?

- Who is responsible for finding and undertaking first stage merger assessments?

- Do you have a comprehensive way to audit potential merger targets — beyond the obvious profit and people issues?

- Do you need to employ external consultants to help with some or all of the change programmes? If so, do you have a clear brief for what is required?

- Do the external consultants have the necessary credibility and compatibility with your partners and staff — as well as the right expertise and experience?

- Are you in danger of overloading your firm with too many change initiatives? If so, prioritise on the most important and focus.

- Regularly take time out to review what you and the firm have achieved and celebrate your successes. Take the time to thank your partners and staff for their hard work and dedication.

Case study: Coda Studios — change management

Nothing is quite as big a change as leaving a large corporate organisation to set up on your own — and then grow your new enterprise at an astonishing but sustainable pace. So it is with great pleasure that I can tell you about Coda Studios — a firm of architects in Sheffield. You will recognise that many of the things at Coda — whether it is clear thinking about the market and its needs, a disciplined approach to setting targets and developing strategies for entering new markets, a considered approach to people management or a constant commitment to profile raising, marketing and relationship development — are mentioned in this book at various places. It is good to see how, when these things are done with enthusiasm and energy, you can generate such fantastic results.

David Cross left a leading architectural practice five years ago to set up on his own — with a desire to avoid the hierarchies of large organisations and do things "his own way". A quick visit to the website (Coda Studios — The art and science of buildings) — which has music, a rainbow coloured index to services and project visuals and animated staff profiles, which include Nelly the dog — shows that the difference from other small practices is more than skin deep. After five years, his practice is going from strength to strength — with 20 people and an ambition to double in size over the next five years. How did he do it?

Early good advice — manage your contacts

I remember at university we had a guest lecturer from WSP and the advice he gave was that we should have a Filofax or some sort of log where we could keep a note of all the contacts we made. Some of the students felt that this was a bit mercenary, but it seemed like good common sense to me — to keep track of who you met and where and their interests — and then to remain in contact. When I was employed I was often asked to do little jobs on the side — and I did and I kept in touch with people. When I was asked to undertake a substantial piece of work — but declined due to a lack of time — I was offered a small monthly retainer for a year and it seemed like an ideal opportunity to set up on my own.

Small beginnings

Initially, I was undertaking small domestic projects — a house extension here and another one there and then it was three, then four and five. Then some new builds. Then I got a few fairly decent commercial jobs. These were often from firms of estate agents and I made an effort to keep in touch with them — just doing the usual stuff like having lunch and dinner, phoning them up and asking how they are doing — even the occasional game of golf. And, of course, I made sure to refer work back to them whenever we could.

First premises

When I persuaded Mark, the other director, to join me we needed office space. My father had bought a building — it had a hairdresser below and an aerobic studio above in Barnsley. There were just two of us in 750 sq ft but from the outset we thought longer term and had a reception desk there and a meeting room and a server-based network to support 10 workstations. Our previous employers had experienced problems with IT, so we were mindful not to have similar difficulties. But it was strange at first — the two of us in all this space with the infrastructure in place for a larger outfit.

Understanding the market

We knew what the market wanted and we undercut the prices quite a bit. We would break down the service we provided into the component parts so clients knew exactly what they were paying for and could choose how much support they purchased. In reality, by selecting most packages they probably ended up paying a similar amount to what they would from other firms — but at least they had transparency.

Raising the profile

Our marketing activity started almost by accident. I bumped into a colleague at The Barnsley Eye when walking the dog. I agreed to write a monthly "Ask the architect" column, where I answered readers' questions on extensions or building homes. It was fantastic exposure and it generated a lot of domestic leads for us. Then we worked with another colleague at *Pad* and *Square Foot* where we bought press pages — we paid for advertising space but used it for articles about projects we were working on. This raised our profile so we ensured that our entries in the three Yell directories in the area were a little larger than our competitors and contained good links to our website.

Innovative website

We worked with a graphic designer to develop a site that was not the usual stuffy, grey and minimalist style often associated with architects. Our brief to him was "corporate funk"! We decided early on that we would put all of our projects on the site — not just the sexy, high profile ones. By providing visuals of all our projects, we could demonstrate that we had a great deal of experience in all manner of work.

Taking on more staff

We had thought that it would be two years before we took on further staff but, in reality, we have hired another person every three months. We made sure that we worked with people for a couple of months before we took them on — so we knew that they were good architects who worked hard and quickly. We have been careful to maintain the relaxed atmosphere and friendly policies to avoid the mistakes of larger corporates. For example, we take all our staff skiing every year and we help them feel empowered by giving them responsibility for their own projects. We pay a monthly bonus and organise plenty of dinners and days out.

Word of mouth
And how did we move from the domestic market into more commercial work? "Well, it was mainly word of mouth. To support all the exposure in the newspapers and magazines, I went to just about every event that I could — awards, RICS events etc. I had a friend at Capita who used to invite me to breakfast and other meetings where I would be among all sorts of business people and those in the construction industry and I networked like mad.

Moving to bigger premises
In our third year, we felt we needed to move to larger premises. We were doing well and wanted a home that had more presence and communicated surety and establishment and that we could brand the Coda way. So we moved into an impressive building in Sheffield.

Managing growth
After sustained growth last year we realised that we might need some help to get to the next stage of our development. After all, we are trained as architects not as business managers. One of our engineering friends had met a consultant called Bernard, who we invited in for a chat. He started off by interviewing all our staff to check that we were doing everything that he thought we should. Thankfully, he was able to reassure us that staff were happy although some of the staff indicated that they would like some guidance on what to do in certain situations, so we developed a staff handbook — the Coda Conduct — which is colourful, bold and well designed.

We also documented some of our policies, such as for maternity benefits. There was some work here too on formalising our chain of command a little — but we have preserved the relaxed and informal feel of the place. We also introduced our team talks where each month we tackle a different topic — sometimes it is a review of the year or the month, other times a topical issue such as the credit crunch and sometimes we invite in a client to give a talk on the basis of "The trouble with architects is ..." so that we keep in touch with what clients do and do not like about our service. As a result of the work we did with staff, we were able to obtain our Investors in People accreditation, which was an important step towards our next strategic goal.

Next, they interviewed some of our clients. Again, we were reassured that we were on track although there were some useful suggestions about our facilities — while we were keen to operate a "no secrets" open office space some of our clients wanted private meeting rooms. We also developed an online customer call centre — where any client could email us if there was any aspect of our service that they wanted to change — I don't think any other architects practice does that.

We do have a business plan — that we wrote ourselves. It outlines the main targets and strategies — for example, what we want to achieve by 2008 and how we plan to do this.

Weathering the storm

To be honest, I do not think we would have done anything any differently. Our guiding principal is "Always be fair". There is no point in ripping people off — it is a small market and word gets around quickly. At the end of the day, we only have our integrity. We have weathered the storm of the credit crunch so far. We did lose a couple of people, but actually they were people that we took on when we were super busy and probably would not have done so in normal circumstances. We have used the slowdown — our turnover is down by around 30% — to take stock and reflect on the past and plan for the future. We have taken time to look at how we are structured, what we are doing and how much we are spending and on what and have managed to reduce our monthly expenditure by about 25%. It has also given us time to work on the next stage of our strategy — developing public sector work — so we have been getting closer to the universities, local authorities and various development quangos. We are now well positioned to move further in this direction.

Case study: RICS

Sean Tompkins is chief operating officer of RICS. In this role, he is responsible for day-to-day organisational performance and for leading the organisation's delivery of its corporate objectives. Prior to joining RICS, Sean was director of marketing for Prudential plc.

Don't forget, please stay in touch ...

As a marketer by background, I have often smiled when I hear that organisations are undergoing some form of "change management".

Why? Well first, because it usually involves a tidy little sum being spent on redirected business effort/consulting advice and second, because to me this is a good indicator that business thinking has been allowed to lose touch with how the marketplace has evolved. This state usually means that either the organisation's tentacles of management information and intelligence have not detected the shift or corporate procrastination has not allowed for the early indicators of change to be responded to anywhere quickly enough. What usually then follows is a period of rapid realignment. This realignment programme puts the business back onto the course it should not have deviated from in the first place if it had ensured it was fully in touch with how the market was changing or decisions to respond to market changes had been taken much earlier.

So why do organisations allow themselves to deviate from their natural course of change?

In my experience, one of the main reasons for this is that they become introverted and fail to spot the everyday signals that stare out of the markets they operate in. Businesses have a daily wealth of information that shows how the pulse of both the organisation and market is beating. Yet it seems that often they do not use it, preferring to rely too much on instinct and previous experience/ learnings to set the course. For most business a blend of information, instinct and experience can help guide solid direction.

The RICS

The RICS (Royal Institution of Chartered Surveyors) is the largest worldwide organisation for professionals operating in the property, land and construction arenas. It was founded in 1868 and over a number of years has awarded chartered titles to more than 100,000 professionals operating out of 140 countries. It operates worldwide and has its headquarters in London. Many of the things it has achieved over many decades owe testament to how it has read the market.

However, about five years ago it recognised a need for dramatic "change" in order to continue to remain relevant in what was a rapidly globalising marketplace. A number of factors had taken hold:

- Members (customers/stakeholders) were less than satisfied with performance.
- Employers of the profession were less than engaged but had a high demand for new talent. The demand for talent was there, yet the image of the profession meant that it was not always competing well against some other professions.
- Property was becoming an asset class in its own right and global investment flows were creating significant market shifts. Employers who had previously recruited chartered surveyors were becoming significant global brands. Investors had amazing choices and on the back of this, standards were gaining increased importance and attention escalated by high-profile failures in other professional fields.

Hindsight is a wonderful thing, but what is now evident is that RICS, at this time, did not have all of the market information and intelligence to hand and some of its thinking and decision making was therefore too centred on the past and "what was", as opposed to the future and "what will be". Subsequently, and to support my argument in the beginning of this case study, RICS could have lessened the scale of change necessary to get back on course had it invested more widely in intelligence and information gathering.

Five years on, what is now different in RICS:

- **Customer complaints**. The transition has been from previously viewing complaints as an interruption to daily life to now viewing them as welcome feedback on how the organisation can continually improve and do better.

- **Market insight**. Getting closer, through developing mutually valuable relationships, with the major stakeholders in the marketplace (employers, opinion formers, policymakers etc). Seeing market movement in advance and gradually realigning propositions and services accordingly.

- **Member (customer) survey**. For us, understanding how our customers/ stakeholders feel, what they see is valuable about what we do, where they see a need to enhance what we do or what we offer. Over the past four years, we have developed significant insight into how our diverse stakeholder base responds to what we do and what it values most.

- **Brand equity/stakeholder audit**. Not an easy thing to measure. For us, our profile, influence and credibility on a worldwide stage enhances the value of our brand. We now conduct regular reviews of wider stakeholder opinion (eg, young future professionals, policymakers) to continue to check the health and future well-being of our brand.

- **Staff survey**. Every six months our staff offer up their views on a range of topics, usually insightfully pointing out the obvious improvements that would yield even better performance.

- **Competitor benchmarking and assessment**.

- **Performance reporting**. Providing an easy format for everyone involved in the running of the organisation to digest performance and look forward to the next steps.

All in all, a transition that has enabled a more confident organisation to emerge, where decisions and future direction, including how the organisation is structured, can be informed by a good dose of external and internal information. Nothing in life will ever be perfect or exact, but you can give yourself a fighting change of getting close at times.

This is rather a simple picture of a case in point and I have taken great liberties in simplifying it for the purposes of looking at how reaching outside of your organisation and investing in staying in touch can mitigate, over time, the extent to which leadership needs to inflict significant change on a business. Perhaps we might one day reflect, that those businesses that seem to rarely change are actually constantly changing and keeping in touch with how their markets are evolving — sadly, they are uninteresting stories, yet probably the true icons of business success.

Appendix 1

Proforma
marketing plan

Executive introduction/ summary	
Where are we now?	
External market analysis	
— Trends, economic, technology, social, political and regulatory changes affecting future client needs	
— Geography (including international developments)	
— Key markets and sectors of interest	
— Competitors	
— Key clients	
— Key prospects	
— Key referrers	
— Key sources	
— Other information required?	

Internal analysis	
— Markets/segments	
— Services/Products	
— Pricing	
— Staff and skills	
— Infrastructure and systems	
— Past and present marketing and business development	
Strengths	
Weaknesses/challenges	
Opportunities	
Threats	
Key issues	

Where do we want to be?

Financial objectives	
Markets/services objectives	
Reputation objectives	
Other objectives	

How will we get there?

Main strategies:	
— Key services	
— Key markets/segments	
— Branding and positioning	
— Key messages	
— Pricing	

Internal development (Recruitment, training, knowledge development etc)	
Existing client development	
New client development	
Referrer development	
Database development	
Research and marketing information systems	

Communications, promotions and profile raising

— Websites, blogs and search engine optimisation	
— Media relations and publicity	
— Publications, brochures and newsletters	
— Signage, reception, meeting rooms and sales/lettings boards	
— Advertising (online and traditional)	
— Key associations/networking	
— Direct and e-marketing	
— Exhibitions and conferences	
— Seminars and workshops	
— Hospitality, receptions and events	
— Key client programmes	
— Referrer programmes	
— Sales strategies	
— Internal communication (team meetings, intranet etc)	
International issues	
Key campaigns and programmes	
Budgets and time	
Monitoring progress and results	

Action plan	
January 2009	
February 2009	
March 2009	
April 2009	
May 2009	
June 2009	
July 2009	
August 2009	
September 2009	
October 2009	
November 2009	
December 2009	

Note: The contents of chapter 4 explain how to complete the various analyses in order for you to complete a strategic marketing plan. The tactical material is covered in chapters 5, 6 and 7 on tactical marketing, selling and clientology.

References

I have a library of thousands of books — and reviewed a good number of them — but just a few of them are mentioned in this book. I have listed some of the others that I have found to be either great introductions to subjects or excellent sources for surveyors and agents with a desire to learn about some of the ideas I have touched on.

Assertiveness at work A practical guide to handling awkward situations Back, Ken and Back, Kate, McGraw Hill (1999)

Best practice creativity, Cook, Peter, Gower (1998)

Beyond Certainty — the changing world of organisations, Handy, Charles, Hutchinson (1998)

Body language at work, Furnham, Adrian, CIPD (1999) The management shaper series is excellent.

Building strong brands Aaker, David A, Simon and Schuster (2002)

Business skills for general practice surveyors, Imber, Austen, EG Books (2005)

Coaching and mentoring — practical methods to improve learning, Parsloe, Eric and Wray, Monika, Kogan Page (2000)

Coaching for performance, Whitmore, John, Nicholas Brealey (1996)

Conceptual selling, Miller, Robert B and Heiman, Stephen E, Warner Books (1987)

Creating new clients — marketing and selling professional services, Walker, Kevin, Ferguson, Cliff and Denvir, Paul, The PACE Partnership (2006) There are a number of other books written by these authors considering other aspects of the business development cycle.

Creating powerful brands in consumer, service and industrial markets, De Chernatony, Leslie and McDonald, Malcolm, Butterworth Heinemann (1998)

Dealing with difficult bosses, Bramson, Robert, Nicholas Brearley (1993)

Dealing with difficult people, Lilley, Roy, Kogan Page (2002)

Developing knowledge based client relationships — leadership in professional services, Dawson, Ross, Butterworth Heinemann (2005)

Dynamic Practice Development — Selling skills and techniques for the professions, Tasso, Kim, Thorogood (2003) Originally published in 1999, it also has an extensive reference section devoted to books on selling.

Empathy selling Golis, Christopher, Kogan Page (1991)

Effective networking for professional success Hart, Rupert, Kogan Page (1996)

Emotional Intelligence, Goleman, Daniel, Bloomsbury (2005) The first and, in my opinion, the best book on the subject.

Even more offensive marketing, Davidson, Hugh, Penguin Business (1997) The sequel to *Offensive marketing*.

Getting to yes — negotiating an agreement without giving in, Fisher, Roger and Ury, William, Random House (1999)

Good to great, Collins, Jim, Random House (2001)

How to win friends and influence people, Carnegie, Dale, Vermilion (2007)

Human resource management, Torrington, Derek, Hall, Laura and Taylor, Stephen, FT Prentice Hall (2007) This is one of the main books on HR used for professional examinations.

Influence: The psychology of persuasion, Cialdini, Robert B, Harpur Business (2007) One of my favourite books of all time.

Influencing — skills and techniques for business success, Dent, Fiona Elsa and Brent, Mike, Palgrave, (2006)

Introducing NLP, Knight, Sue, CIPD (1999) The management shaper series is excellent.

Key account management, Cheverton, Peter, Kogan Page (2004)

Key account management — learning from supplier and customer perspectives, McDonald, Malcolm and Rogers, Beth, Butterworth Heinemann (1998)

Kotler on marketing — How to create, win and dominate markets, Kotler, Philip, Free Press (2001)

Leadership and the one minute manager, Blanchard, Ken, Harper Collins (2000)

Listening skills, Mackay, Ian, CIPD (1998) The management shaper series is excellent.

Making sense of law firms — Strategy, structure and ownership, Mayson, Stephen, Blackstone Press (1997) Stephen Mayson is a leading expert in law firm management and many of his ideas can be applied to property partnerships.

Making successful presentations Forsyth, Patrick, Sheldon Business Books (1995)

Malcolm McDonald on marketing planning — understanding marketing plans and strategy, McDonald, Malcolm, Kogan Page (2008) This is an easy-to-read introduction for non-marketers.

Managing key clients — securing the future of the professional services firm, Walker, Kevin, Ferguson, Cliff and Denvir, Paul, Continuum (2000)

Managing the professional service firm, Maister, David, Free Press (2003) All of David Maister's books are worth a read — they are all written especially for the professions.

Marketing plans. How to prepare them: how to use them, McDonald, Malcolm, Butterworth Heinemann (2002) Malcolm McDonald is a true marketing guru and has written many books. This is the main book on marketing plans.

Marketing the professional services firm — Applying the principles and the science of marketing to the professions, Young, Laurie, Wiley (2005) This is the standard text for the Cambridge College of Marketing's Chartered Institute of Marketing diploma for marketing in the professions. Sometimes he teaches in the next classroom to me.

Media relations in property, Norwood, Graham and Tasso, Kim, EG Books (2006) This book focuses exclusively on media relations in the residential and commercial property sectors. Graham is a leading residential property journalist.

Motivating people Heller, Robert Dorling Kindersley (1998)

Negotiate — The art of winning, Mills, Harry A, BCA (1991)

Negotiating, persuading and influencing, Fowler, Alan, CIPD (1998) The management shaper series is excellent.

NLP — The new technology of achievement, Andrea, Steve and Faulkner, Charles, Nicholas Brealey (1996)

People watching — How to take control, Coleman, Vernon, Blue Books (1995)

Practice Management guidelines for surveyors, Kennie, Tom RICS Books (2003) A short but essential guide — it also covers financial planning this RICS Guidance Note is mentioned in the introduction.

Presenting with impact — A masterclass McConnell, John Biddles (2004)

Presenting with power — Captivate, motivate, inspire and persuade, McConnon, Shay, Howtobooks (2006)

Professional service firm 50, Peters, Tom, Random House (1999)

Selling with integrity: Reinventing sales through collaboration, respect and servicing, Morgan, Sarah Drew, Berrett Koehler (1999)

Smarter selling, Dugdale, Keith and Lambert, David, Prentice Hall (2007) This was on the bestseller list — and is excellent.

Snakes and ladders for property professionals: How to be a smooth operator in the property and construction industry, Kay, Frances, EG Books (2007)

Spin Selling, Rackman, Neil, Gower (1995)

Starting and developing a surveying business: For sole traders, Imber, Austen, EG Books (2005) This is for those who have not quite created a partnership yet.

Strategic brand management, Elliott, Richard and Percy, Larry, Oxford University Press (2006)

Strategic brand management Keller, Kevin, Lane Prentice Hall (2008)

Techniques of structured problem solving, Van Grundy Jr, Arthur B, Van Nostrand, Reinhold (1993)

The shorter MBA — A practical approach to the key business skills Pearson, Barrie and Thomas, Neil, Profile Books (2004)

The back of the napkin — solving problems and selling ideas with pictures, Roam, Dan, Penguin Group (2008) This book explains how to use drawings to solve problems and present complex ideas in a simple but effective way.

The complete idiot's guide to leadership, DuBrin, Andrew J, Alpha Books (1998)

The complete idiot's guide to project management, Baker, Sunny and Baker, Kim, Alpha Books (2000)

The definitive guide to direct and interactive marketing, Stone, Merlin, Bond, Alison and Blake, Elizabeth, FT Prentice Hall (2003)

The new strategic selling, Heiman, Stephen E, Miller, Robert and Tuleia, Tad, Business Plus (2005) One of the best professional sales systems.

The one minute manager, Blanchard, Ken and Johnson, Spencer, Harper Collins (2000)

The power of persuasion — how we're bought and sold, Levine, Robert, Oneworld (2007)

The seven habits of highly successful people, Covey, Stephen R, Franklin Covey (2004)

The trusted adviser, Maister, David, Free Press (2002)

Time to think, Kline, Nancy, Cassell Illustrated (1999)

Transitions: Making sense of life's changes, Bridges, William, Nicholas Brealey (2004)

True Professionalism — The courage to care about your people, your clients and your career, Maister, David, Simon & Schuster (2001)

Understanding brands: Creating success Cheverton, Peter, Kogan Page (Sunday Times Creating Success) (2000) A simple introduction.

Wally Olins on Brand Olins, Wally Thames & Hudson (2004)

Win that pitch, Bell, Quentin, Kogan Page (1993)

Index